Seoho Jung

Fabrication and Operation of Fast, Ultra-low-power Gas Sensors with Carbon Nanotubes

Seoho Jung

Fabrication and Operation of Fast, Ultra-low-power Gas Sensors with Carbon Nanotubes

Hartung-Gorre Verlag Konstanz

Reprint of Diss. ETH No. 30560

Scientific Reports on Micro and Nanosystems **Volume 40**

edited by Prof. Dr. Christofer Hierold
ETH Zürich
Micro and Nanosystems

Cover images:
Front:

- **Left:** Robotic assembly of an individual, suspended carbon nanotube
- **Middle:** Raman image of the gas sensor
- **Right:** Suspended carbon nanotubes grown between cantilevers

Back:

- **Left:** Output characteristics showing negative differential conductance
- **Middle:** Schematic drawing of self-heating-assisted sensor recovery
- **Right:** Transient sensor signals

Bibliographic information published by Die Deutsche Nationalbibliothek

Die Deutsche Nationalbibliothek lists this publication in the Deutsche Nationalbibliografie; detailed bibliographic data is available in the internet at http://dnb.dnb.de.

First edition 2025
HARTUNG-GORRE VERLAG, KONSTANZ
http://www.hartung-gorre.de

ISSN 2566-7769
ISBN 978-3-86628-836-2

Diss. ETH No. 30560

Fabrication and Operation of Fast, Ultra-low-power Gas Sensors with Carbon Nanotubes

A thesis submitted to attain the degree of
DOCTOR OF SCIENCES
(Dr. sc. ETH Zurich)

presented by

Seoho Jung
MSc ETH in Energy Science and Technology, ETH Zurich

born 10 April 1991

accepted on the recommendation of

Prof. Dr. Christofer Hierold, examiner
Dr. Cosmin Roman, co-examiner
Prof. Dr. Michel Calame, co-examiner
Prof. Dr. Joan Daniel Prades, co-examiner

2024

Abstract

The exceptional properties of carbon nanotubes (CNTs), including their high mechanical strength, electron mobility, and thermal conductivity, make them promising candidates for various applications. One possible application is the sensing of nitrogen dioxide (NO_2), a toxic pollutant with significant implications for air quality monitoring, pollution control, and medical applications. Existing sensor technologies often fail to meet the stringent requirements for power consumption, shelf life, readout speed, and limit of detection (LOD) necessary for mobile applications. CNT-based gas sensors offer substantial potential to address these limitations, yet several challenges must be overcome to utilize their full potential.

One major technological hurdle is the high-volume fabrication of CNT devices, in particular that of suspended CNT devices, which demonstrate several advantages as sensors. This thesis presents an approach to address this challenge, focusing on the scalable batch growth of suspended CNTs and their automated assembly into functional devices. Through systematic multi-parameter optimization and the development of large-scale growth substrates, the optimal growth conditions for CNTs were identified, leading to a significant increase in the number of individual CNTs available for device integration. An automated nanoassembly machine was employed to transfer pre-selected CNTs from growth substrates to device substrates. The process flow designed around automated robotic assembly demonstrates a production rate – mechanical transfer rate for carbon nanotubes – of approximately 46 CNTs per hour, with a 70 % yield of electrically active devices. The assembled devices were characterized electrically to understand the performance of devices fabricated by this method.

Contact resistance at the CNT-electrode interface is another critical factor influencing device performance. This thesis demonstrates that electrode etching with argon ion immediately prior to nanotube transfer effectively removes all adsorbates from palladium or gold electrode surface and, as a result, consistently enables clean CNT-electrode contact. When combined with post-transfer thermal annealing, this pre-transfer treatment reduces the median ON-resistance (R_{on}) of CNT devices by an order of magnitude and the interquartile range by more than two orders of magnitude. This is crucial for the self-heating functionality of suspended CNT gas sensors.

Lastly, the thesis demonstrates the application of suspended CNTs in fast and ultra-low-power NO_2 sensors. By leveraging self-heating for rapid gas desorption and employing a nonlinear transient analysis method, the sensors achieved a signal recovery time of 1 minute and readout time of 1-5 minutes at a peak power of approximately 5.6 μW. With a readout time of 5 minutes, the limit of detection of the sensor was 9 ppb NO_2 in synthetic air with 40 % relative humidity. These performance improvements were reproduced with additional devices, which were assembled both manually and automatically.

Zusammenfassung

Die außergewöhnlichen Eigenschaften von Kohlenstoffnanoröhren (CNTs), einschließlich ihrer hohen mechanischen Festigkeit, Elektronenmobilität und Wärmeleitfähigkeit, machen sie zu vielversprechenden Kandidaten für verschiedene Anwendungen. Eine mögliche Anwendung ist die Detektion von Stickstoffdioxid (NO_2), einem giftigen Schadstoff mit erheblichen Auswirkungen auf die überwachung der Luftqualität, die Kontrolle der Verschmutzung und medizinische Anwendungen. Bestehende Sensortechnologien können oft die strengen Anforderungen an Energieverbrauch, Haltbarkeit, Auslesegeschwindigkeit und Nachweisgrenze (LOD) für mobile Anwendungen nicht erfüllen. CNT-basierte Gassensoren bieten erhebliches Potenzial zur Bewältigung dieser Einschränkungen, aber mehrere Herausforderungen müssen überwunden werden, um das volle Potenzial der CNT-basierten Gassensoren zu verwenden.

Eine der großen technischen Herausfoderungen ist die Massenfertigung von CNT-Bauelementen, insbesondere von Bauelementen mit suspendierten CNTs, die als Sensoren mehrere Vorteile bieten. Diese Arbeit stellt ein Verfahren vor, zudem das Batch-Wachstum von suspendierten CNTs und automatische Transfers von den CNTs zu funktionalen Bauelementen gehören. Durch systematische Multi-Parameter-Optimierung und die Entwicklung von großflächigen Wachstumssubstraten wurden die optimalen Wachstumsbedingungen für CNTs identifiziert, die zu einer signifikanten Erhöhung der Anzahl einzelner CNTs führten, die für den Trocken-Transfer verfügbar sind. Eine automatisierte Nanoassembly-Maschine wurde eingesetzt, um ausgewählte CNTs von Wachstumssubstraten auf Bauelemente zu transferieren. Der Prozessablauf, der auf einem automatisierten robotergestützten Transfer basiert, demonstriert eine durchschnittliche Transfersrate von 46 CNTs pro Stunde. Die Bauelemente wurden elektrisch charakterisiert, um deren Leistung zu verstehen, und 70 % von den Bauelementen waren elektrisch aktiv.

Ein weiterer wichtiger Faktor, der die leistung von den Bauelementen beeinflusst, ist der Kontaktwiderstand an dem CNT-Elektroden-Interface. Diese Arbeit zeigt, dass das Ätzen der Elektroden mit Argonionen direkt vor dem Transfer der CNT die Adsorbate auf der Elektrodenoberfläche effektiv entfernt und dadurch durchgehend einen sauberen

CNT-Elektroden-Kontakt ermöglicht. In Kombination mit einer thermischen Nachbehandlung nach dem Transfer reduziert diese Vorbehandlung den Median von den ON-Widerständen (R_{on}) der CNT-Bauelemente um eine Größenordnung und den Interquartilbereich um mehr als zwei Größenordnungen. Dies ist entscheidend für die Selbstheizfunktionalität von suspendierten CNTs in Gassensoren.

Abschließend demonstriert die Arbeit die Anwendung von suspendierten CNTs in schnellen und energieeffizienten NO_2-Sensoren. Durch Nutzung der Selbstheizung für eine schnelle Gasdesorption und Anwendung einer nichtlinearen Transientenanalyse-Methode erreichten die Sensoren eine Signalerholungszeit von 1 Minute und eine Auslesezeit von 1-5 Minuten bei einer Spitzenleistung von etwa 5,6 μW. Mit einer Auslesezeit von 5 Minuten betrug die Nachweisgrenze des Sensors 9 ppb NO_2 in synthetischer Luft mit 40 % relativer Feuchtigkeit. Diese Leistungsverbesserungen wurden mit zusätzlichen Bauelementen reproduziert, die sowohl mit manuellen CNT-Transfers als auch mit automatischen CNT-Transfers hergestellt wurden.

Acknowledgements

First, I would like to express my deepest gratitude to my advisor, Prof. Christofer Hierold. His guidance and encouragement have significantly shaped my research and way of thinking. His profound insights into the research area and his support inside and outside of the laboratory have been instrumental in shaping this work.

Having Dr. Cosmin Roman as my co-supervisor and co-examiner has been a great privilege. He was always ready to help me with everything from high-level strategies to bottom-level details. An hour-long conversation with him often saved me days or even weeks of time.

I thank Prof. Michel Calame for co-examining my thesis and leading the Nanoassembly research consortium I was part of. His expertise in nanoelectronics opened my eyes to nanomaterials and technologies beyond my PhD research. I also thank Prof. Joan Daniel Prades for co-examining my thesis and sharing his insights into ultra-low-power gas sensing. His earlier and recent works have been great guidelines I could refer to.

I am also very grateful for the advice and support of Dr. Miroslav Haluska. Much of what I have learned about carbon nanotube growth and Raman spectroscopy during my PhD study was possible through his guidance.

I have been fortunate to share this journey with my colleagues at the Micro and Nanosystems group, who have, without exception, been great colleagues. I would like to thank Morten Vollmann, Cristina Gentili, Ines Kraiem, Dr. Lalit Kumar, Dr. Laura Jenni, Dr. Cancan Blaser, Dr. Sebastian Eberle, Dr. Stefan Nedelcu, Dr. Kishan Thodkar, Max Aue, Dr. Mainak Basu, and Hichem Ben Lakhdar for everything I learned from them about carbon nanotube research. My thanks go further to my colleagues Dr. Signe Vehusheia, Dr. Matthias Dupuch, Dr. Ian Mihailovic, Dr. Katrina Klösel, Dr. Johannes Weichart, Lucrezia Maini, Florin Püntener, and Sira Bielefeldt for the collaboration. Outside of the Micro and Nanosystems group, I would like to thank all members of the Nanoassembly consortium for the successful collaboration. These include the members of Transport at Nanoscale Interfaces Laboratory (Empa, Prof. Calame), Institute for Machine Tools and Manufacturing (ETH

Zurich, Prof. Wegener), Machine Learning and Optimization Laboratory (EPFL, Prof. Jaggi), and inspire AG (Dr. Weikert).

I have enjoyed a privilege to continue working on my research in the form of a company. My colleagues at Chiral Nano AG – Dr. Natanael Lanz, Dr. Andre Butzerin, Dr. Maria El Abbassi, Dr. Luca Ornago, Marko Nikolic, and Jeroen Zandbergen – have been of enormous help in my last sprint toward the finish line. Collaboration with Frederik van Veen and Prof. Mickael Perrin has always been an interesting experience.

I have had the pleasure of supervising several students during my studies. I would like to thank Jiacheng Jiang, Julian Radziwill, Simon Zimmer, Roland Bockholt, Ricardo Stirnimann, David Seda, and Kristian Ivkovic for their contributions and enthusiasm. Their hard work laid much of the groundwork for this thesis.

A special thanks to Manuela Kägi for managing the group and always being there to help with administrative matters. Her assistance with numerous administrative needs has been greatly appreciated.

The cleanroom facilities at BRNC (IBM Research) and FIRST-CLA are maintained by dedicated staff members, and I am grateful for their support. I would particularly like to thank Dr. Ute Drechsler, Dr. Ronald Grundbacher, and Donat Scheiwiller for their excellent help and support.

I am deeply grateful to my family, extended family, and my partner Lena for their kindness and understanding. I owe them much for the wonderful time I have had during and beyond my academic journey.

Finally, I would like to acknowledge the financial support from Strategic Focus Area - Advanced Manufacturing program of ETH Board, which provided the opportunity to collaborate with groups across ETH Domain. The support from SNSF FLAG-ERA CONVERGENCE (20FE-1170224) is also appreciated.

List of Symbols

Symbol	Description	Unit
I_d	Drain current	A
K_{ads}	Adsorption coefficient	min^{-1} ppm^{-1}
K_{des}	Desorption coefficient	min^{-1}
$L_{contact}$	Contact width	m
L_{gate}	Gate distance	m
L_{sd}	Source-drain pitch	m
p	partial pressure	Pa
R_{CNT}	Channel resistance	Ω
$R_{contact}$	Contact resistance	Ω
R_{device}	Total device resistance	Ω
R_{on}	ON-resistance	Ω
R_Q	Quantum resistance	Ω
SS	Subthreshold swing	mV/dec
V_{sd}	Source-drain bias	V
V_{gs}	Gate bias	V
V_{th}	Threshold voltage	V
θ	Surface coverage	None

List of Abbreviations

Abbreviation	Description
CNT	Carbon nanotube
SEM	Scanning electron microscopy
AFM	Atomic force microscopy
XPS	X-ray photoelectron spectroscopy
CVD	Chemical vapor deposition
RIE	Reactive ion etching
ALD	Atomic layer deposition
FET	Field-effect transistor
SGS	Small-gap semiconducting
RBM	Radial breathing mode
NW	Nanowire
NP	Nanoparticle
SH	Self-heating
UV	Ultraviolet
LOD	Limit of detection
R.H.	Relative humidity
SNR	Signal-to-noise ratio

Contents

List of Figures

List of Tables

1 Introduction

A color version of this chapter is available at https://doi.org/10.3929/ethz-b-000702474.

1.1 Ultra-low-power gas sensors for air quality monitoring

World Health Organization (WHO) estimates that, every year, more than 6 million deaths are attributable to the detrimental effects of air pollution [1]. Among numerous pollutants, six pollutants draw particular attention as especially harmful threats to human health: $PM_{2.5}$, PM_{10}, O_3, SO_2, CO, and NO_2.

The concentration guidelines for these pollutants for a 24-hour time window are as follows:

Table 1.1: 24-hour concentration guidelines for the six most actively monitored air pollutants. Conversion to ppb (parts per billion) or ppm (parts per million) values follows the method of [2].

Pollutant	$PM_{2.5}$	PM_{10}	O_3	SO_2	CO	NO_2
Actual limit	75 $\mu g/m^3$	150 $\mu g/m^3$	82 ppb (8-hour)	48 ppb	6 ppm	64 ppb
Target limit	15 $\mu g/m^3$	45 $\mu g/m^3$	51 ppb (8-hour)	15 ppb	4 ppm	13 ppb

This thesis focuses on nitrogen dioxide (NO_2), a toxic gas usually generated by combustion processes. As noted in Table 1.1, like O_3 and SO_2, it can pose a threat to human health at concentrations as low as a few tens of ppb (parts per billion) [3]. Monitoring NO_2 levels at this concentration level is thus important for ensuring compliance with environmental regulations and protecting public health. Therefore, the monitoring of NO_2 requires sensing systems with a very low limit of detection (LOD). In addition, due to the prevalence of combustion processes – a major source of NO_2 – in various industries [4], there is an increasing need to monitor this gas with mobile systems. Semiconducting metal-oxide (MOX) sensors, the current industry standard, are hindered by selectivity issues, short lifetime, and high power [5]. Electrochemical amperometric sensors have an advantage in power, but are typically bulky and short-lived [6, 7].

In the past decades, several NO_2 sensors based on nanostructures including metal oxide (MOX) nanowires/nanoparticles [8–10], low-dimensional materials [11–13], and metal-organic frameworks (MOFs) [14] have demonstrated great potential as an alternative, especially in terms of LOD and power. However, most of the devices in the literature do not demonstrate sufficiently short readout and recovery times, which should be on the order of a few minutes for practical NO_2 monitoring in the field [15].

In this thesis, the terms recovery time, response time, and readout time are used as follows. Recovery time is defined as the duration required for the sensor signal to return to its baseline level. Response time refers to the time needed for the sensor signal to reach a specified percentage of its final equilibrium value upon exposure to the target gas. Readout time is the time required to determine the analyte concentration.

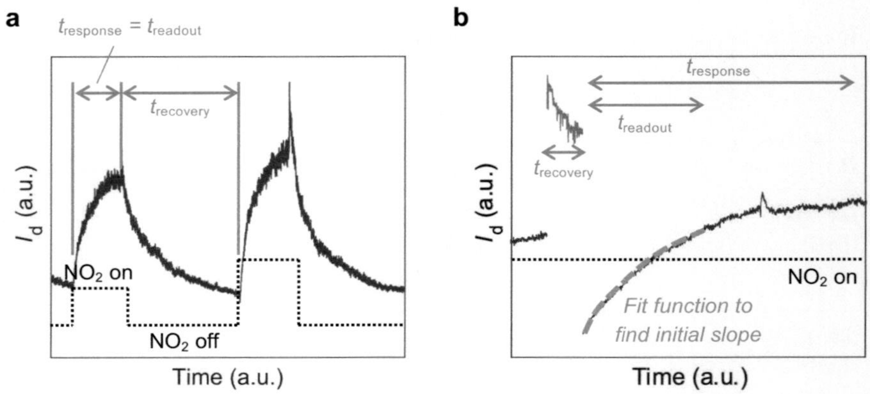

Figure 1.1: Schematic illustrations of recovery time ($t_{recovery}$), response time ($t_{response}$), and readout time ($t_{readout}$) in (a) steady-state sensing and (b) initial slope sensing.

In steady-state sensing, as illustrated in Figure 1.1a, the readout time is the same as the response time because the analyte concentration can only be determined when the sensor signal reaches its final equilibrium value or a specified percentage thereof. t_{90}, the time needed to reach 90 % of the total signal change, is used widely. Similarly, recovery time is often quantified as the time needed for the signal to return to 10% of the signal change upon gas exposure (t_{10}).

In the case of initial slope sensing coupled with assisted sensor recovery (Figure 1.1b), the definitions differ. Initial slope sensing is a type of transient sensing, which is detailed extensively in Chapter 5 of this thesis [16, 17]. In this operation scheme, sensor recovery is accelerated by an external heater or self-heating of the sensing material. Subsequently, the initial slope of the sensor transient after recovery is then used to determine the analyte concentration. In this case, $t_{readout}$ can be shorter than $t_{response}$ as the initial slope can be determined before the sensor signal reaches its steady state.

A sensor's reporting interval, the total time needed to update the sensor's concentration reading, is then given by:

$$\text{Reporting interval} \geq t_{\text{recovery}} + t_{\text{readout}}.$$

In an ideal scenario, the recovery mode and readout (sensing) mode would alternate continuously to achieve the shortest reporting interval.

The requirements for a mobile NO_2 sensor can be summarized as follows:

- **Limit of detection (LOD)**: The LOD should be ideally smaller than the concentration guidelines outlined above – i.e. a few tens of ppb or below.

- **Readout time**: The readout time in the order of a few minutes due to the constantly fluctuating nature of the pollutant concentration [15].

- **Recovery time**: The recovery time should be on the similar range to the readout time, as a full measurement cycle requires both readout and recovery.

- **Energy consumption (power)**: Peak power of a few μW would be desirable for use in battery-powered portable devices, dense sensor networks, or IoT applications.

- **Selectivity**: The sensor needs to be able to selectively detect the target analyte in a gas mixture. Cross-sensitivity to humidity and other gas species should not hinder accurate sensing of the analyte.

- **Fabrication scalability**: For large-scale deployment beyond research, high-volume production should be possible.

- **Long-term stability**: The sensor should endure years of operation in real-life conditions.

1.2 Carbon nanotubes for NO_2 sensing

As noted earlier, recent gas sensor research has actively investigated the use of nanostructures as potential alternatives to traditional sensing materials, and carbon nanotube is certainly one of those nanomaterials. Since Kong *et al.*'s first report on the response of carbon nanotubes to adsorbed NO_2 and NH_3 in 2000 [18], the use of CNTs or carbon nanotube field-effect transistors (CNTFETs) as gas sensors has drawn consistent attention. CNTs possess several properties that make them attractive candidates for gas sensors [19, 20]:

- **High surface-volume ratio**: Carbon nanotubes have all their atoms on the surface, which enable the highest surface-to-volume ratio possible. As a result, almost all the carbon atoms in the nanotube can interact with the analyte gas.

- **Surface-sensitive electron transport**: The delocalized Π clouds that govern charge transport in nanotubes are at the surface. Adsorbing species and electrostatic charges close to the nanotube can therefore directly influence charge transport.

- **Structural stability**: C-C bonds are very strong ($\sim$5.3 eV per bond in benzene). Thus, the nanotube is very stable to repeated temperature cycling and chemical exposure.

Among different architectures of CNTFETs, suspended CNTFET, in which a carbon nanotube is suspended between source and drain electrodes, as opposed to being attached to the substrate (i.e. substrate-bound CNTFET), offers several advantages as a gas sensor:

- **Suppressed hysteresis**: Suspended nanotube channel is isolated from oxides or charge traps [21, 22], and thus shows almost hysteresis-free transfer characteristics (I_d-V_{gs}) [23].

- **Reduced drift and noise**: The absence of charge traps also suppresses baseline drift and charge noise [24].

- **Possibility of fast, low-power sensor recovery**: CNT-based NO_2 sensors suffer from slow recovery at room temperature, which takes >12 hours in case of NO_2. However, an individual, suspended CNT can Joule-heat significantly at a very low power, and this self-heating effect can be used to accelerate the recovery of sensor signal. For example, Chiakkdi *et al.* show (almost) complete sensor recovery in 10 minutes at 3 μW [25].

For these reasons, this work focuses on devices with individual, suspended nanotubes as NO_2 sensors.

1.2.1 Challenges in carbon nanotube-based NO_2 sensing

However, there still remain a few important challenges on the path to effective application of CNT-based NO_2 sensors, and some of them are listed below:

- **Electrode-nanotube contact**: This is especially relevant to the low-power sensor recovery enabled by nanotube self-heating, which is described above. In order for the suspended nanotube channel to heat up significantly, the electrical contact between the electrode and the nanotube has to be very good because, in case of poor contact with high contact resistance ($R_{contact}$), the voltage drop at contacts prevents the suspended nanotube channel from heating up. This is why the realization of self-heated CNT gas sensors has been difficult and rare, characterized by a very low device yield [20].

- **Scalable fabrication**: Suspending an individual nanotube between source and drain electrodes is not an easy task, and direct nanotube growth [26–28] and mechanical transfer [29, 30] are the two techniques that are used for most of the suspended FETs in the literature. Direct growth typically requires very high (>800 °C) growth temperature, which is detrimental for existing circuit components on the device. Mechanical transfer can serve as an ultraclean, room-temperature alternative integration method, but is a manual and slow process. Automation and acceleration of this mechanical transfer technique can make the fabrication process more scalable.

- **Slow sensor response**: The conventional steady-state sensing scheme requires a long readout time ($>$2 hours), whereas the readout time for a practical application should not exceed a few hours [15]. Among the proposed solutions in the literature, transient sensing could reduce the readout time to a few minutes [17]. However, this requires power-intensive (few milliwatts) microheaters for sensor recovery, nullifying CNT's advantage in low power.

The following chapter (Chapter 2) will repeat some of these points and explain where these discussions stand in the context of nanostructure-based gas sensing research.

2 State of the Art

This chapter aims to review the state of the art of nanostructure-based gas sensors and nanomaterial integration methods. This discussion is supplemented by shorter and more focused literature reviews of other specific topics in Chapters 3, 4, and 5.

The first half of the literature review (Section 2.1) focuses on nanostructure-based gas sensors. Section 2.1.1 starts with an introduction of commercial gas sensors and their limitations in mobile, low-power sensing applications, which lead to the motivation behind the ample literature that investigates nanostructure-based gas sensors as an alternative to commercial systems. First, we visit sensors based on 1D nanomaterials like nanowires (Section 2.1.2) and nanotubes (Section 2.1.3). Before moving onto other nanostructures, there is an in-depth discussion about the self-heating operation of suspended 1D nanomaterials (Section 2.1.4). Then, the literature review covers gas sensors based on other nanostructures like nanoparticles and 2D nanomaterials (Section 2.1.5). The section concludes with a focused summary of nanostructure-based NO_2 sensors in Section 2.1.6.

The other half of the literature review (Section 2.2) is dedicated to the integration of nanomaterials into functional devices. Section 2.2.1 discusses non-deterministic integration methods for nanowires, carbon nanotubes, and 2D materials. It is followed by a discussion of deterministic integration methods for the same materials (Section 2.2.2). Section 2.2.3 summarizes and compares different robotic assembly systems, as robotic assembly attracts more and more interest as a promising technique for integration of individual nanomaterials.

A color version of this chapter is available at https://doi.org/10.3929/ethz-b-000702474.

2.1 Nanostructure-based gas sensors

2.1.1 Commercial gas sensors

Traditional, large-scale air quality monitoring infrastructures are based on bulky and costly technologies such as gas chromatography [31], mass spectrometry [32], and ion mobility spectrometry [33]. The mobility of these systems is severely limited by their substantial size (in the range of a few tens of centimeters) and weight (in the range of a few kilograms). In addition, they typically require installation and extensive maintenance by specialized personnel. Therefore, while these systems are essential for highly precise and selective measurements, the gas sensor industry has invested heavily in the development of miniaturized gas sensors.

The push towards miniaturization in gas sensors has resulted in several commercial sensors. Market examples include electrochemical sensors such as AlphaSense's NO_2-B43F [34] and metal oxide (MOX) sensors such as Bosch Sensortec's 4-in-1 (temperature, humidity, pressure, and gas) sensor BME680 [35], ams OSRAM's low-power volatile organic compounds (VOCs) sensor CCS811 [36], Sensirion's multi-pixel gas sensor SGP41 [37], and ScioSense's multi-gas sensor ENS160 [38]. For a more extensive overview of commercial gas sensors, the reader is kindly directed to Nedelcu (2022) [2].

However, as the summary in Table 2.1 suggests, the main drawbacks of these sensors are usually found in power and limit of detection (LOD). These sensors generally require power in the mW range to demonstrate limits of detection below the concentration thresholds defined by governmental organizations.

This is why the industry and academia further pursue their research into gas sensors based on low-dimensional (0D, 1D, and 2D) nanostructures. These efforts engineer metal oxides, which have been widely used in bulk forms like films, into nanoparticles (0D) or nanowires (1D) for better sensing performance and lower power. In addition, they also explore the use of nanomaterials like carbon nanotube (CNT), graphene variants, and transition-metal dichalcogenides (TMDs) as sensing materials. The following sections present an overview of these technologies. As information about the industry's research and development is not easily accessible to the public, the review focuses mainly on academic literature. Since NO_2 sensors are the most relevant to this thesis, they remain the core of the literature review. However, sensors for other analytes are also

Table 2.1: Summary of selected commercial gas sensors. Values not directly reported have been estimated from information available in spec sheets. Unless specified otherwise, LOD is determined with 3σ confidence or the lowest concentration level, at which sensor response is demonstrated. MOX = Metal oxide, N/A = Not available.

Company	Type	LOD	Readout time	Power
AlphaSense [34]	Electrochemical	20 ppb NO_2	~1 min	100 μW
Bosch Sensortec [35]	MOX	N/A	1-120 sec	200 μW - 25 mW
ams OSRAM [36]	MOX	N/A	N/A	1-50 mW
Sensirion [37]	MOX	10-20 ppb NO_2	<5 min	~10 mW

introduced occasionally if the technology serves to highlight the advances in nanostructure-based gas sensing.

2.1.2 1D nanomaterials: nanowires

Nanowires made up of metal oxides (MOX) and metal oxide semiconductors (MOS) are the most actively investigated nanostructures as gas sensors. Their well-defined geometry, single-crystalline nature, and small diameters all contribute to a large surface-to-volume ratio, which is expected to provide a high sensitivity of gas sensors. Empirical research and theoretical simulations consistently emphasize the importance of reducing crystal size and increasing surface area in maximizing the sensitivity of conductometric gas sensors [39]. It is the reactions occurring at the surface and interfaces that are pivotal in the adsorption of gases and the subsequent sensing performance. An increased surface-to-volume ratio substantially elevates the proportion of surface atoms of metal oxides, making more sites available for interaction with gas molecules [40].

Networks of nanowires

The majority of nanowire-based gas sensor studies use networks or arrays of nanowires, rather than individual nanowires. This is largely due to the fabrication complexity of individual nanowires. However, networks of nanowires already show advantages of using 1D nanostructures, as opposed to films, in gas sensing.

Artzi-Gerlitz *et al.* [41] clearly exemplifies these advantages by comparing the sensing response of the hollow nanotubes of WO_3 to that of thin films of the same material. The hollow nanotubes formed by sol-gel deposition in an anodic aluminum oxide membrane show sensor responses – defined as $\Delta R/R_0$ – to 20 ppm of NO_2 that are 2 to 3 orders of magnitude higher than that of thin films. Furthermore, the response time of the nanotube-based sensor is approximately 1 minute, which is an order of magnitude improvement. The authors attribute this significantly improved sensing performance to the nanotube's high surface-to-volume ratio. One expected limitation of this sensor is power. Although the study does not explicitly report the device's power, the operating temperature of 200 °C for the sensor surface of 1 cm^2 is likely to require milliwatts of power. In addition, the reported LOD of 200 ppb is a bit higher than what is shown by more recent technologies. Despite these limitations, this study serves to clearly highlight the advantages of using 1D nanostructures for gas sensing by direct comparison with films.

Jiang *et al.* [42] use SnO_2 nanocrystalline tubes to demonstrate an NO_2 sensor with an impressive LOD of 10 ppb. These structures, formed by electrospinning, feature a high surface-to-volume ratio, which, when paired with their inherent porosity, is believed to facilitate enhanced gas diffusion. This contributes to relatively quick response times on the order of several tens of seconds. Despite these advantages, the long-term stability of these sensors remains an issue, as a few consecutive tests already suggest a reduction in sensitivity and a compromise of response speed, which are ascribed to the absorption of humidity on the sensor's surface.

Along with SnO_2, ZnO is among the most widely investigated materials. ZnO nanowires draw particularly strong interest, due to the easy and straightforward synthesis methods available for for high-quality, single-crystalline 1D ZnO nanostructures [43].

For example, in Chen *et al.*'s work [44], tree-branched ZnO nanowires synthesized in-situ on a micro-hotplate area with a small diameter of 20 nm demonstrate resoluble detection to 5 ppb H_2S. This performance can be compared to that of traditional ZnO nanowires that typically measure around 50 nm in diameter. The study shows that the 50-nm nanowires only respond to H_2S concentrations exceeding 50 ppb. Although this work is not about NO_2 sensors, it is still another example of the nano-sizing effect contributing to higher chemical activity at surface.

Silicon nanowire-based devices also make low-LOD NO_2 sensors. Cuscuna *et al.* [45] employs plasma-enhanced chemical vapor deposition (PECVD) to directly synthesize a network of silicon nanowires on interdigitated electrodes, demonstrating the potential to detect NO_2 concentrations as low as 10 ppb at room temperature. The fabrication is completely based on the silicon technology platform. The underlying mechanism of this device is the formation of nitric acid through the reaction between NO_2 and water molecules, which leads to a strong interaction of amined silicon nanowires even with very low NO_2 amounts. As a result, such a low LOD is attainable only under humid conditions and it rises drastically to approximately 1 ppm in a dry atmosphere.

Another example of a low-LOD NO_2 sensor is demonstrated with a highly ordered silicon nanowire array on a flexible substrate [46]. The aligned array of silicon nanowires is fabricated with a method known as superlattice nanowire pattern transfer with precise control over the nanowire dimensions and spacing. The nanowires formed on a silion-on-insulator (SOI) wafer are dry-transferred onto a flexible substrate by etching the buried oxide and releasing the wires for transfer. Contacts and device structures are then established using conventional microfabrication processes. Chemical functionalization of the nanowires with a variety of silanes allows the sensor array to distinguish between different gases via multi-component analysis, as demonstrated with acetone and hexane. While this method is interesting for its potential adaptability to a wide range of target gases and substrates, the sensor's response and recovery times in the range of tens of minutes can hinder practical application. In addition, the sensor quickly shows a drift of 1 % after just 10 minutes of operation.

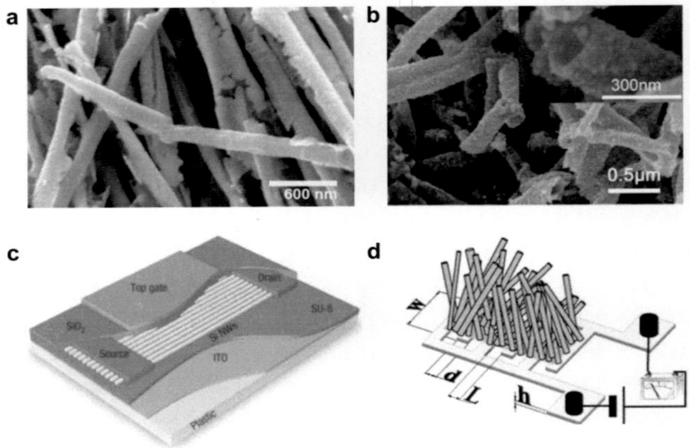

Figure 2.1: NO_2 sensors comprising networks of MOX/MOS nanowires. (a) SEM image of hollow WO_3 nanotubes formed by sol-gel deposition. [41]. (b) SEM image of SnO_2 nanocrystalline tubes formed by electrospinning [42]. (c) Schematic illustration of an ordered silicon nanowire array fabricated with superlattice nanowire pattern transfer [46]. (d) Schematic illustration of a network of silicon nanowires synthesized by PECVD [45]. Reproduced with permission from [41,42,45,46]. © Elsevier S.A. 2009, Royal Society of Chemistry 2012, Springer Nature 2007, AIP Publishing 2012.

Individual nanowires

The works introduced so far rely on networks or arrays of nanowires. While these structures already demonstrate the advantages of nanowires, the ideal architecture that can possibly benefit the most is a sensor based on an *individual* nanowire. The further simplification and miniaturization of the structure – from networks of nanowires to individual nanowires – result in several advantages, with the most notable being lower power.

Chinh *et al.* systematically shows this in a study [9], in which individual, multiple, networked SnO_2 nanowire sensors are fabricated and compared. Sensors based on an individual nanowire show advantages over those comprising multiple or networked nanowires, including better gas response, shorter response and recovery times, and lower power. The study especially focuses on the possibility of heating up an individual nanowire via self-heating. Under self-heating conditions at a bias of 25 V, the sensor demonstrates a significant response to NO_2 gas concentrations as low as 1 ppm and short response times of approximately 8 seconds and recovery times of about 25 seconds, all at only 35 μW of power. Such response of the sensor is comparable to that of externally heated sensors at 150 °C, but the power of 35 μW is far less than what advanced external heaters, even those designed with MEMS technology, would require to heat a sensor to 150 °C. The self-heating operation of 1D nanostructures is discussed in further detail in Section 2.1.4.

Chinh *et al.* emphasize that, despite the superior gas sensing performance of individual nanowire sensors, their fabrication is significantly more complicated than networks of nanowires and is therefore characterized by a low yield. This fabrication difficulty presumably explains why individual nanowire-based sensors are not more common in the literature despite their advantages.

Zhang *et al.* report an LOD of 20 ppb NO_2 with an individual In_2O_3 nanowire at ambient temperature [47], which is claimed to be an order-of-magnitude improvement compared to the earlier studies using films. Like [9], this work also compares an individual In_2O_3 nanowire with multiple In_2O_3 nanowires. In this case, the multiple-nanowire sensor configuration is shown to be more reliable and sensitive, achieving a lower detection limit of 5 ppb NO_2. In addition, this configuration demonstrates better selectivity over NH_3, O_2, CO, and H_2. However, such relative advantages of multiple nanowires are obtained at the cost of a significantly (~20 times) longer response time of 1000 seconds and are attributed to the averaging mechanism in the multiple-nanowire sensors.

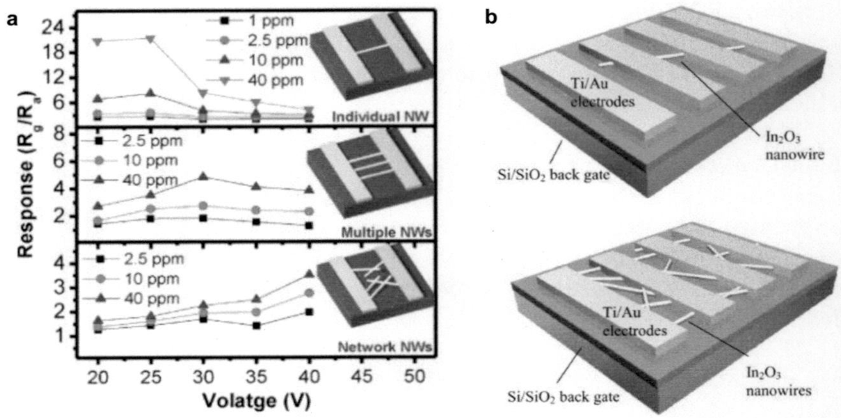

Figure 2.2: Comparison of NO_2 sensors consisting of individual/networked nanowires. (a) A sensor based on an individual SnO_2 nanowire shows significantly shorter response and recovery times and lower power than a sensor based on networked SnO_2 nanowires [9]. (b) A sensor based on an individual In_2O_3 nanowire is 20 times faster in response, despite a slightly higher limit of detection [47]. Reproduced with permission from [9, 47]. © Elsevier S.A. 2014, American Chemical Society 2004.

2.1.3 1D nanomaterials: carbon nanotubes

Carbon nanotube (CNT) is the ultimate form of 1D nanostructure. The nanowires discussed so far have a diameter in the range of a few tens or hundreds of nanometers, and this corresponds to tens or hundreds of layers of atoms. In contrast, carbon nanotube is a hollow tube made up of an atomically thin graphene sheet, and an individual, single-walled carbon nanotube has a diameter of 1-2 nm. In this 1D system, charge carriers move presumably along the tube's axis, and adsorbing species and electrostatic charges close to the nanotube can directly influence charge transport. These changes in charge transport can then be measured as signals when a nanotube is used in a sensing system.

Following Iijima's historic discovery of CNTs [48], Kong *et al.* first reported the electrical behavior changes of CNT field-effect transistors (CNTFETs) upon gas exposure, setting the stage for CNT-based gas sensor research [18]. These sensors, characterized by their nanometric scale and remarkable electrical properties, promise exceptional sensitivity and low power requirements.

The promise of novel gas sensors that can potentially overcome the limitations of the above-mentioned MOX/MOS nanowire-based sensors has led to ample literature on CNTFET-based gas sensors. Research progressively addresses a wide range of topics including response speed, recovery, power, and so on. This section explores the evolution of different device architectures for CNT-based NO_2 sensor technology, with a focus on non-functionalized nanotubes. Theoretical discussions about carbon nanotube's band structure, non-idealities such as hysteresis and noise, gas sensing mechanisms, and kinetics are not included in this chapter. The reader is directed to Chikkadi (2013) [19] for an extensive review of these topics. For a discussion about functionalization methods and gas sensors based on functionalized nanotubes, a summary in Eberle (2019) [20] is recommended.

Networked nanotubes vs. individual nanotubes

The electrical characteristics and gas sensing performance of CNT-based sensors are intricately tied to the device architectures – and the fabrication techniques used to realize the architecture. As a result, extensive research has investigated into various architectures. As is the case for MOX/MOS nanowire-based sensors, devices incorporating individual carbon nanotubes are considered more complicated to fabricate than devices with networked nanotubes. As a result, the majority of the studies, especially early works in the field, use networked nanotubes – with a famous exception of Kong *et al.*'s very first report [18], which is based on an individual nanotube.

The networks of nanotubes are mostly realized by solution-based assembly [49–52] or direct growth of nanotubes on substrates [53].

Li *et al.*, one of the first studies using solution-based assembly, deposit a network of carbon nanotubes on interdigitated gold (Au) electrodes by a simple solution casting process [49]. Utilizing multiple nanotubes helps mitigate nanotube-to-nanotube variations in a similar way to the 'averaging' mechanism for In_2O_3 nanowires described by [47]. This way, they report an LOD of 44 ppb NO_2, highlighting that non-functionalized nanotubes can make a highly sensitive sensor. Between their measurements, the sensor surface is cleaned with 2-hour ultraviolet (UV) light exposure because unassisted desorption of NO_2 is very slow at room temperature.

An array of nanotubes aligned by dielectrophoresis also shows fast response to ~1 ppm NO_2 [51]. Xu *et al.* uses CNTs dispersed in a sodium dodecyl sulfate (SDS) surfactant solution and observes the successful directional alignment of CNTs between electrodes due to the electrophoretic force in a nonuniform electric field. The aligned array shows ~100 % increase in conductance when exposed to ~1 ppm NO_2 at room temperature. However, as in [49], the recovery of CNT surface after gas adsorption relies on UV irradiation for 10 minutes.

Ammu *et al.* report thin films of CNT bundles on a flexible paper substrate [50]. This sensor device can detect concentrations as low as approximately 125 ppb NO_2, with quick response and recovery times of a few minutes. The study emphasizes the importance of the choice of substrate material in the fabrication of organic thin-film sensors by comparing paper-based and PET-substrate-based sensors. In case of PET-substrate-based sensors, the sensor does not recover from gas exposure without external heating or UV irradiation.

Jeon *et al.* explore the impact of device dimensions on gas sensing capabilities by utilizing a network of pre-separated 99.9 % semiconducting CNTs [52]. They demonstrate that adjusting the nanotube network's dimensions – length and width – within their CNTFET design can help to determine the optimal gate bias for H_2 and NO_2 sensing. For NO_2 sensing, the device shows the highest sensitivity in the subthreshold region, so making CNT networks narrower and thereby increasing the ON-OFF ratio is desirable. The limitation of this study is that, due to their focus on device design, the gas measurements are performed at elevated temperatures and at concentrations well above the levels relevant to practical application.

Unlike the four studies introduced above, Cantalini *et al.* uses PECVD-grown nanotubes to offer insights into how slight alterations in catalytic particle sizes used in the synthesis of carbon nanotubes (CNTs) can significantly influence their electrical properties and, as a result, the sensing performance of the final NO_2 sensors [53]. Adjusting the thickness of the Ni catalyst film from 3 nm to 5 nm increases the nanotubes' size (diameter) and makes the network more oriented. This results in different responses to NO_2 at different temperatures between sensors. Both sensors show response to NO_2 concentrations as low as 10 ppb at 165 °C, yet sensors with smaller CNTs exhibit higher signals but more baseline drift compared to those with larger CNTs. This study underscores the importance of CNT properties in the behaviors of the final sensors made out of the CNTs.

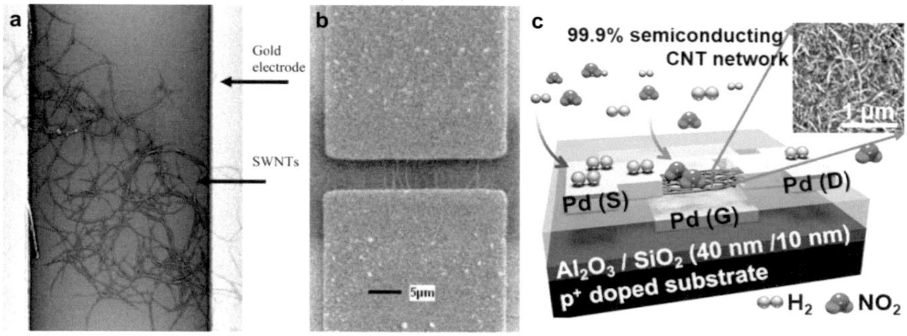

Figure 2.3: NO_2 sensors based on networked carbon nanotubes. (a) Carbon nanotube network formed on interdigitated gold electrodes by a solution casting process [49]. (b) A sparse array of carbon nanotubes aligned by dielectrophoresis [51]. (c) A network of pre-separated 99.9 % semiconducting carbon nanotubes [52]. Reproduced with permission from [49, 51, 52]. © American Chemical Society 2003, Taylor & Francis 2012, AIP Publishing 2017.

Investigating gas sensors based on individual CNTs may offer deeper insights and, at the same time, potential to fully leverage CNTs' highly sensitive electronic transport and low power. However, fabrication of a device with exactly one of this extremely small nanomaterial remains a big challenge in the field. Therefore, the number of studies reporting gas sensors based on individual CNTs is still very limited.

One of the early studies using an individual nanotube as a gas sensor is Chang *et al.* [54]. This work mainly focuses on demonstrating the sensor recovery by gate voltage pulses after exposure to NO_2, not on characterizing the sensor performance. As a result, the gas exposure is done with a very high concentration of 300 ppm NO_2, and the sensor performance metrics are not analyzed.

Chen *et al.* suggests that in-situ cleaning of the nanotube with UV during gas measurements significantly improves the sensor's sensitivity [55, 56]. For a device with an individual nanotube, the LOD is determined to be 44 ppb NO_2. They also address gas selectivity by introducing a gate bias. However, the response and recovery times are in the range of a few tens of minutes, and the power of the continuous UV irradiation, which is likely in the milliwatt range or above, is not specified.

A recent work by Sacco *et al.* aims to combine the advantages of individual nanotubes and networked nanotubes by fabricating a sensor consisting of a few individual nanotubes connecting interdigitated electrodes [57]. They suggest the 100-ppb LOD of the sensor is the outcome of combining the high sensitivity of individual sensors and the large sensing area and channel current of networked nanotubes. Although not very high, this LOD is still above the regulatory concentration limits for NO_2.

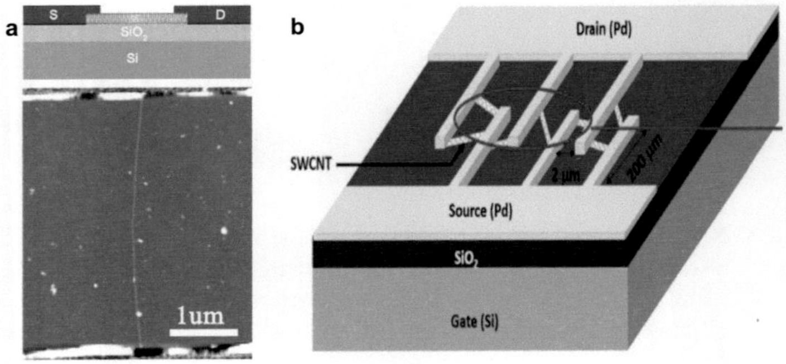

Figure 2.4: NO_2 sensors based on individual carbon nanotubes. (a) Top-contacted, individual CNT grown by CVD [54]. (b) A few individual nanotubes connecting interdigitated electrodes [57]. Reproduced with permission from [54, 57]. © IOP Publishing 2007, Elsevier S.A. 2020.

Despite the scarce literature on individual nanotube-based gas sensors, the Micro and Nanosystems Group at ETH Zurich has consistently advanced research in this direction.

A brief summary of the recent history of the group's research in carbon nanotube-based gas sensor starts with a work by Mattmann *et al.*, which presents a contact-passivated NO_2 sensor with an LOD as low as 100 ppb NO_2 and an improved recovery time from several hours to about 30 min [58]. Here, the recovery is accelerated by heating the sensor at 110 °C. The sensor's response despite the passivation of metal contacts with leak-tight Al_2O_3 suggests that there is a sensing mechanism based on the interaction between the gas molecules and the open CNT channel.

Mattmann *et al.*'s next work focuses on suppressing the gate hysteresis of the sensor with pulsed gate sweep strategies [59]. In this pulsed gate sweep, the device current I_{sd} is not recorded over V_g sweeping continuously going from one end of the sweep window to the other, Instead, V_g is incremented in steps, returning to 0 V after each pulse. Distinctions are made between continuously increasing or decreasing pulses (p++ pulse) and pulses that alternate polarity (p+- mode), effectively minimizing or eliminating hysteresis. This reduction in hysteresis occurs regardless of pulse duration for alternating pulses due to discharging of charge traps. Optimal results were observed at the shortest pulse duration, $\tau_{pulse} = 0.5$ ms. These observations are consistent in dry air and upon exposure to NO_2 as low as 100 ppb.

Individual, suspended nanotubes

All devices introduced thus far – both nanowire-based in 2.1.2 and nanotube-based in 2.1.3 – are substrate-bound devices, in which the nanowires or nanotubes are attached to the substrate. However, there is another architecture of devices, in which 1D nanostructures are *suspended* between source and drain contacts. Figure 2.5 illustrates these two types of devices with carbon nanotube as an example of 1D nanostructure.

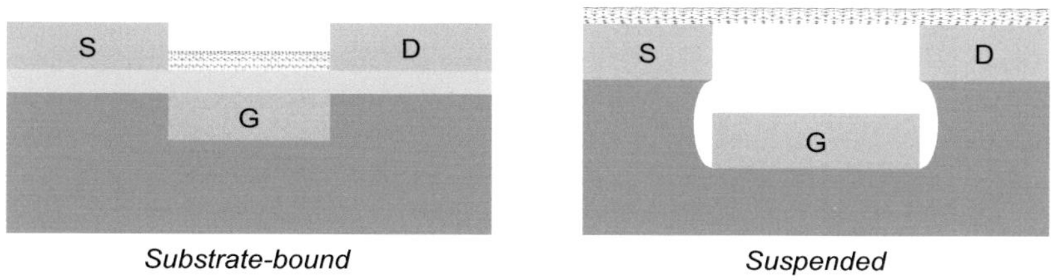

Figure 2.5: Schematic illustrations of a substrate-bound CNT field-effect transistor (left) and a suspended CNT field-effect transistor (right).

The concept of suspended CNTFETs, especially the ones with *individual*, suspended nanotubes, is important to note at this point because this architecture is shown to be another effective way to suppress hysteresis almost completely. Muoth *et al.* grow a suspended carbon nanotube directly between the source and drain electrodes of a polysilicon substrate and contact the nanotube by subsequent metal evaporation using a movable shadow mask [23]. The nanotube integrated into the device this way is ultraclean, as it is not exposed to any processing chemical after growth. The shadow masking technique used here does not limit the choice of the contact metal, which is typically not compatible with the high temperatures required for nanotube growth in *growth-last* fabrication methods. The isolation of the nanotube channel from the substrate reduces the gate hysteresis to only a few mV even in humid, room-temperature environments, and this significant change is attributed to the removal of charge traps. It has to be noted, however, that this fabrication process is complex, time-consuming, and random as it involves chemical vapor deposition (CVD) growth of nanotubes, which is probabilistic by nature, and manual manipulation of a shadow mask with micrometer-level precision. As a result, the yield of devices with an individual, suspended nanotube is fairly low around 2 %.

Muoth *et al.* proposes an alternative fabrication method that can overcome the probabilistic nature of the CVD growth process – mechanical transfer of suspended nanotubes [29]. This method grows suspended nanotubes on a separate fork-like substrate and then mechanically transfers them onto a device substrate, preserving the ultraclean state of the nanotube. Devices made this way demonstrate hysteresis-free transfer characteristics (I_d-V_{gs}).

These two studies do not report any gas measurements, but the suspended device architecture and the mechanical transfer method lead to the realization of NO_2 sensors with significantly reduced hysteresis in the subsequent works of the group introduced below.

Chikkadi *et al.* compare the behavior of suspended CNTFET gas sensors to that of conventional substrate-bound devices [24]. The suspended CNTFET devices do not show observable baseline drift under constant gate biases, while the signal-to-noise ratio upon exposure to 1 ppm NO_2 is 9 times higher than substrate-bound devices. Suspended CNTFET devices show an extrapolated LOD of 32 ppb and a power below 1 mW [60]. However, the response time is ~2 hours and the recovery time is ~12 hours.

The next work of Chikkadi *et al.* proposes a breakthrough solution for the slow recovery problem [25]. Suspended CNTs, due to their minimal heat dissipation into substrates, can undergo substantial Joule heating when applied a high source-drain bias voltage V_{sd}, functioning like an ideal heating filament [61]. Chikkadi *et al.* induce this self-heating effect in dry air after after exposing the sensor to 200-700 ppb of NO_2. After each self-heating pulse, the sensor baseline is recovered, implying that the gas molecules have been desorbed from the nanotube surface. The longer the self-heating pulse, the higher the recovery rate. The signal recovery of 93 % is achieved after just 10 minutes of self-heating at a bias $V_{sd} = 1.3$ V. Even more impressive is the fact that the peak power during the self-heating is only 3 μW.

The reduction of sensor recovery time from 12 hours to 10 minutes is a significant improvement, but the long sensor response time still remains a challenge for CNT-based NO_2 sensors. Therefore, Satterthwaite *et al.* propose a novel signal readout method based on transient signals [17]. They recover the sensor by regularly heating the sensor with an external microheater and monitor the initial slope of the sensor's transient response during the re-adsorption of gas molecules on the reset surface. The initial slope can be utilized to quickly determine the concentration of the analyte in 7-10 minutes, a process inherently faster than waiting for a steady-state signal that takes 2 hours. However, the use of an external microheater requires a power in the milliwatt range, nullifying suspended CNT's typical advantage in low power. In addition, even with the use of the heater, the sensor recovery time is longer than an hour.

2.1.4 Self-heating of suspended 1D nanostructures

As discussed earlier, 1D nanostructures can heat up to high temperatures even with small amounts of electrical power. When used in gas sensor technology, this self-heating effect makes it possible to reduce the power of sensors down to the microwatt regime. In the case of conductometric gas sensors, this is a factor of 100-1000 improvement from the state-of-the-art microsensors. In case of nanowires, the most accepted models describe the self-heating effect as a factor that controls the transport balance between chemisorption of gas species at the surface and desorption of products (Figure 2.6a) [62]. Typically, the temperatures, at which adsorption and desorption are optimally balanced, are in the range of a few hundreds of degrees celsius. On the other hand, carbon nanotubes are gas-sensitive at room temperature without functionalization and need heating only to accelerate the desorption of gas molecules (Figure 2.6b).

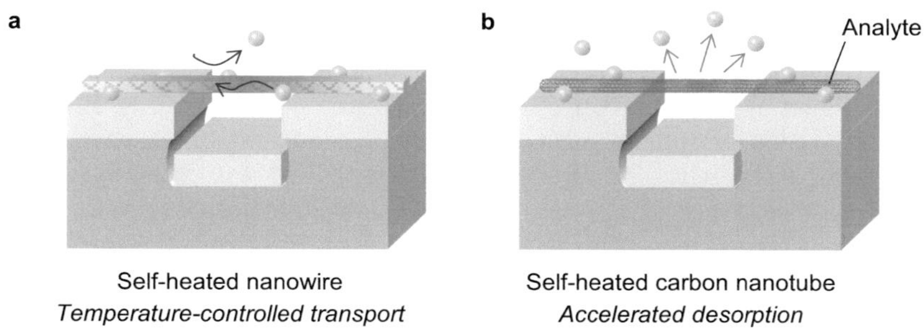

Figure 2.6: (a) Self-heating effect in a suspended nanowire affects the energetic barriers for chemisorption of gas species at the surface and desorption of products. (b) A suspended carbon nanotube is gas-sensitive at room temperature, and the self-heating effect accelerates the desorption of gas molecules.

Although the self-heating effect has been demonstrated with substrate-bound architectures too (e.g. SnO_2 film [63]), individual, suspended 1D nanostructures are the architecture that can benefit the most from the self-heating effect. This is, as discussed earlier, due to the absence of heat dissipation into the substrate.

Kawano *et al.* are the first in the literature to report the application of self-heating effect in sensing [64]. They show a conductometric sensor based on an individual nanotube, the resistance of which changes depending on both the bias conditions (i.e. the amount of energy dissipated) and the pressure of the atmosphere surrounding the nanotube. They attribute these changes to the different thermal losses by convection to different gas pressures.

Not much later, Strelcov *et al.* [65,66] and Prades *et al.* [8,67] report the first chemical gas sensors in a series of studies using individual SnO_2 nanowires. Especially, [8] proves that different electrical biases lead to gas responses fully equivalent to those obtained at different temperatures controlled with an external heater, showing that the responses are the result of the sensor activation by temperature. The work also demonstrates SnO_2 nanowire-based sensors for NO_2 that can operate at 20 μW. They demonstrate a thermal response time of a few ms and a gas response time of 5 minutes.

Fabrega *et al.* present a very comprehensive overview of self-heated, low-power sensors based on 1D nanostructures reported until 2017 [62]. This section complements Fabrega *et al.*'s review by introducing more recent advances from 2017 onward, with a focus on individual, suspended 1D nanostructures.

Seo *et al.* present a H_2 gas sensor composed of a 200-nm-thick suspended carbon nanowire decorated with Pd nanoparticles of varying sizes, with an aim to detect H_2 gas at room temperature [68]. The novel architecture featuring carbon nanowires suspended and decorated with Pd nanoparticles allows high sensitivity, broad detection range, and fast sensor recovery through ultra-low-power self-heating. Especially, the recovery is complete in 5 seconds at 30 μW. The fabrication process leverages techniques like carbon-MEMS and electrodeposition, and the sensor sensitivity changes with the deposition thickness of Pd nanoparticles. The integration of nanowires with Pd nanoparticles of different sizes helps to achieve fast and complete recovery of sensor signal with low power.

Another Pd-decorated nanowire sensor is reported by Yun *et al.* [69], who fabricate Pd-decorated silicon nanowires to achieve low power and high stability against environmental noise. The nanowires are produced through standard CMOS processes, with Pd nanoparticles deposited via physical vapor deposition. The nanowires are then suspended by etching away the buried oxide layer. The Pd-decorated nanowires need less than 1 mW for Joule heating, demonstrate fast H_2 response within 3 seconds without sensitivity loss – a common issue in other Pd-based sensors. The suspended nanowires show a similar sensing performance at significantly reduced joule heating power (147 μW) compared to their substrate-bound counterparts

(613 μW), effectively maintaining H_2 sensing accuracy even in humid or carbon monoxide environments.

Prajapati *et al.* use Pt nanowires for H_2 gas detection at room temperature [70]. By electron beam lithography, sputtering, and lift-off processes, sub-100 nm diameter Pt nanowires are fabricated. As the surface of Pt oxidizes during the nanowire self-heating, the nanowires turn into PtO_x/Pt nanowires. This sensor architecture detects hydrogen down to 100 ppm with ultra-low power of approximately 45 nW. Forming an array of oxidized Pt nanowires increases the sensitivity significantly by more than an order of magnitude compared to single nanowire sensors. With an array, the authors demonstrate a detection limit as low as 500 ppb H_2 with short response (25 seconds) and recovery (108 seconds) times.

While the three papers above report H_2 sensors, Lim *et al.* use a ZnO-decorated carbon nanowire as a NO_2 sensor [71]. A ZnO nanowire forest is grown around a suspended carbon nanowire fabricated via carbon-MEMS and hydrothermal methods. This design, leveraging a suspended structure for easier functionalization, significantly increases gas molecules' access to ZnO sensing areas, improving mass transfer and sensitivity. This sensor shows NO_2 sensing capabilities with an LOD of 50 ppb and a linear sensor response in the 50-300 ppb range, outperforming traditional substrate-bound ZnO nanowire sensors.

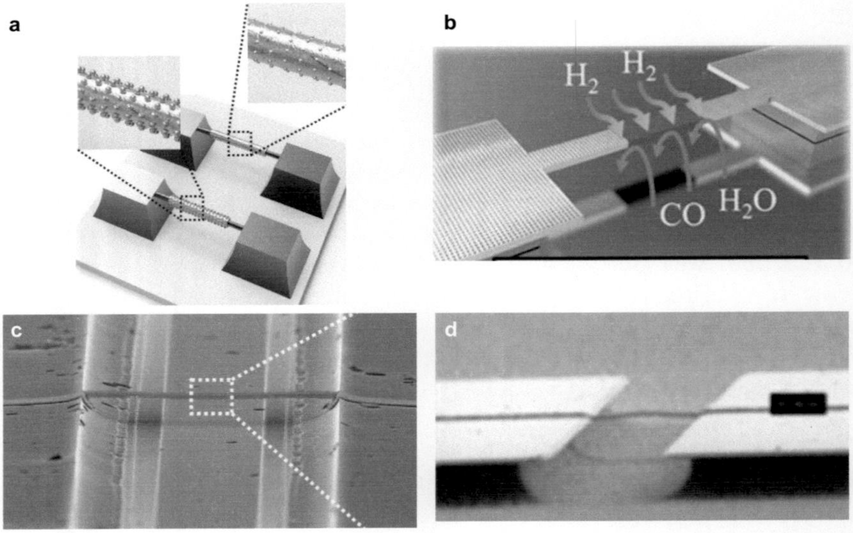

Figure 2.7: Self-heating of individual, suspended nanowires. (a) Pd-decorated carbon nanowire for H_2 sensing [68]. (b) Pd-decorated silicon nanowire for H_2 sensing [69]. (c) ZnO-decorated carbon nanowire for NO_2 sensing [71]. (d) Pt nanowire for H_2 sensing [70]. Reproduced with permission from [68–71]. © Elsevier S.A. 2017, American Chemical Society 2019, Elsevier S.A. 2018.

In case of gas sensors based on an individual, suspended carbon nanotube, important advances have been highlighted in Section 2.1.3. Nanotube self-heating reduces the sensor recovery time from 12 hours to 10 minutes, and the peak power during self-heating is only 3 μW [25]. Another sensor of the same architecture demonstrates an extrapolated LOD of 32 ppb [60]. However, to make these sensors viable for practical application, the slow response to NO_2 (2-3 hours) should be reduced significantly or a faster readout method should be developed as an alternative to the conventional steady-state sensing.

2.1.5 Other nanostructures: nanoparticles and 2D materials

Besides 1D nanostructures, 0D nanostructures like MOX nanoparticles and 2D nanomaterials like graphene are also widely used in gas sensing.

Recent studies with MOX nanoparticles aim to couple the nanoparticles with a low-power activation source such as a micro light plate (μLP) [10] or an LED [72]. In [10], a film of ZnO nanoparticles is combined with a micro light plate (μLP) featuring an InGaN-based LED (455 nm) to realize a low-power NO_2 sensing system. The LED is only a few hundred nanometers away from ZnO. This configuration allows controlled, uniform, high-irradiance photoactivation at a low power. Responses to NO_2 show a bell-like curve with significant detection (20 % response) to 25 ppb at $30\,\mu$W power. If the power is increased to 200 μW, the response grows to 94 %, which yields an extrapolated LOD of 1 ppb. The sensor displays some level of selectivity towards NO_2 over NH_3, CO, and CH_4.

Among 2D nanomaterials, graphene is the closest relative of carbon nanotubes, and offers many of the same advantages as carbon nanotubes. Its electronic transport is highly adsorbate-sensitive [73]. Chemically converted graphene variants such as reduced graphene oxide (rGO) make the material preparation process easier and cheaper, and have been successfully applied for gas sensing [74]. However, the advantage in easy fabrication comes at the cost of compromised sensing performance, as the LOD (~2 ppm NO_2) and the response time (40 minutes) of rGO are well above the desired values. A more recent work reports a graphene-based NO_2 sensor on a transparent, flexible substrate [11], and it shows that the self-heating operation of graphene is not very power-efficient. For its best performance – 7 ppb NO_2 LOD, 10 minutes response time, and 50 minutes recovery time – the sensor requires 14 mW, which is too high for the response and recovery times that are still too long. Therefore, graphene sensors appear to be only be suitable for analytes that do not require short response and recovery times at room temperature.

2.1.6 Summary: nanostructure-based NO_2 sensors

Table 2.2 summarizes some of the most relevant achievements in nanostructure-based NO_2 sensing. Instead of listing all the studies available in the literature, the table aims to to feature the best-in-class values for each group of nanostructures and systems.

The table introduces several impressive advances in the field of nanostructure-based gas sensing. However, at the same time, it highlights the difficulty of developing a sensor that fulfills all the requirements in LOD, readout time, recovery time, and power at the same time.

When it comes to ultra-low-power operation, sensors based on self-heated, suspended 1D nanostructures show clear advantages. Especially, the SnO_2 nanowire sensor by Prades *et al.* [8] demonstrates well-rounded performance metrics. Carbon nanotube-based sensors show further advantages in LOD and power [25, 60]. If the issue with the long sensor readout time can be resolved, carbon nanotube-based sensors can further cement their potential for practical applications.

This thesis builds upon Chikkadi *et al.*'s use of nanotube self-heating effect to further reduce the sensor recovery time to 1 minute. In addition, it proposes a new readout mechanism based on physics-based transient sensing, which shortens the readout time to 5 minutes and demonstrates a LOD as low as 9 ppb. Chapter 5 discusses these results in further detail.

For the sake of comparison with the state of the art, the table already includes data of the sensor that will be presented in Chapter 5 (see "This work" in the last row). One of the notable differences between this work [75] and previous studies using individual, suspended carbon nanotubes (Chikkadi *et al.* [25, 60], Nedelcu *et al.* [76]) lies in the sensor recovery time. Section 5.3.1 suggests that the reduction of recovery time from 10-45 minutes to 1 minute is possible mainly through the choice of optimized gate bias (V_{gs}) and provides a physical explanation that focuses on different sensing regions of a suspended CNT-based gas sensor.

Table 2.2: Comparison of selected nanostructure-based NO_2 sensors. Values not directly reported have been estimated from plots. Unless specified otherwise, LOD is determined with 3σ confidence or the lowest concentration level, at which sensor response is demonstrated. For works that use steady-state sensing, readout/recovery time is the time when 90 %/10 % of the steady-state value is reached. NPs = Nanoparticles, NW = Nanowire, μLP = Micro light plate, SH = Self-heating, UV = Ultraviolet, N/A = Not available.

Work	**Material**	**LOD**	**Activation method**	**Readout time**	**Recovery method**	**Recovery time**	**Peak power**
Casals *et al.* [10]	ZnO NPs	1 ppb (5σ)	LED (μLP)	80 min	None	100-200 min	200 μW
Kim *et al.* [11]	Graphene	7 ppb	SH	10 min	None	50 min	14 mW
McAlpine *et al.* [46]	Networked Si NWs	20 ppb	None	>15 min	None	>15 min	N/A
Zhang *et al.* [47]	Multiple In_2O_3 NWs	5 ppb	None	20 min	UV	20 min	N/A
Prades *et al.* [8]	Individual, suspended SnO_2 NW	100 ppb	SH	2 min	None	2 min	20 μW
Lim *et al.* [71]	Individual, suspended ZnO/C NW	50 ppb	SH	4 min	None	7 min	N/A
Chikkadi *et al.* [25,60]	Individual, suspended CNT	32 ppb	None	120 min	SH	10 min	3 μW
Nedelcu *et al.* [76]	Individual, suspended CNT	23 ppb	None	40 min	SH	45 min	2 μW
This work [75]	Individual, suspended CNT	9 ppb	None	5 min	SH	1 min	6 μW

2.2 Nanomaterial integration methods

The integration of a nanostructure is a crucial part of the fabrication of a functional nanoelectronic device. This section introduces different nanostructure integration techniques, with a focus on the integration of *individual* nanostructures into devices. The techniques can be largely grouped into two categories: non-deterministic methods and deterministic methods. In the context of this section's discussion, the term *deterministic* means placement of an individual nanostructure at a pre-determined spot.

2.2.1 Non-deterministic integration

Nanowires

Section 2.1.2 introduced several MOX or MOS nanowire-based gas sensors. This section presents a more detailed discussion of how the nanowires are synthesized and then integrated into functional electronic devices – not only limited to gas sensors.

The discussion starts with the synthesis or growth of nanowires. Nanowires are typically grown by a vapor-phase method such as a vapor-liquid-solid (VLS) mechanism [77], which is typically performed in a furnace with controlled temperature, pressure, and gas flow rates. In this environment, the source material is heated until it evaporates. This vapor is then transported via a carrier gas – usually a mixture of O_2 and inert gases – to a substrate. The controlled pressure of the inert atmosphere and the gradient of temperature within the furnace allow condensation and nucleation of the nanostructures on the substrate. Catalysts may either be deposited in advance on the substrate or form spontaneously during the process. This method allows precise control over the nanowire growth, resulting in highly pure, structurally uniform, and single crystalline nanowires.

These nanowires are grown in the form of networks or vertically aligned forests, and for device fabrication, they are transferred onto a device substrate by a dry or wet transfer method.

Contact printing method presented by Javey *et al.* is one of direct, dry transfer methods [78]). They demonstrate oriented and patterned assembly of nanowires on various substrates. The growth substrate containing vertical, randomly oriented Ge/Si nanowires is pressed against a patterned device substrate, slid slightly by 1-3 mm, and retracted. The shear forces during this process align the nanowires to a certain extent. Devices and circuits are fabricated on these nanowire arrays using lithography and metallization. They compare 40 FETs based on these arrays and demonstrate small variations in threshold voltage and average ON-current, with a standard deviation of only 15 %. This ability to average variations across multiple nanowire arrays is claimed as one of the most notable advantages of this technique, along with the applicability to a wide range of nanowire materials and device substrates.

In case of wet transfer, nanowires are initially dispersed in solvents like methanol, ethanol, isopropanol, or water at low concentrations. For example, in [79], a diluted ZnO suspension is drop-cast onto a substrate and allowed to dry. The positions and angles of these structures on the substrate are determined by microscopy, and top metallic contacts are formed through a lift-off process or deposition through a shadow mask. This step is then followed by an annealing process to improve the adhesion of structures to the electrode material. ZnO-based photodetector devices are fabricated in this way. Despite its ability to precisely place and orient nanowires, this approach is still characterized by its complexity and low throughput.

Nanowires can also be wet-transferred onto pre-patterned bottom contacts. In this case, the chance of having nanowires accurately aligned between electrodes can be quite low without any alignment mechanism. Therefore, a method like dielectrophoresis can be used to improve the alignment of nanowires [80]. However, the surface of pre-patterned bottom electrodes is easily vulnerable to contamination, so obtaining good electrical contact between the electrodes and nanowires is a challenge. This inconsistency in contact quality leads to irreproducibility and makes this technique less ideal for applications requiring low-resistance contacts.

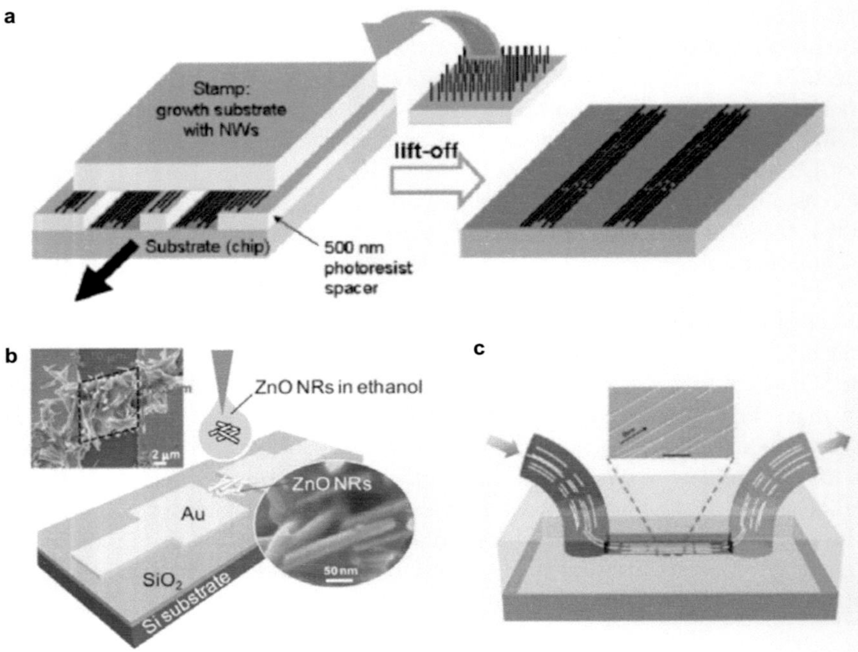

Figure 2.8: Non-deterministic integration of multiple nanowires. (a) Contact printing of Ge/Si nanowires [78]. (b) Drop casting of ZnO nanowires dispersed in ethanol [79]. (c) A parallel array of InP nanowires aligned by a microfluidic flow [81, 82]. Reproduced with permission from [78, 79, 81, 82]. © American Chemical Society 2007, Wiley-VCH GmbH 2013, The American Association for the Advancement of Science 2001, American Chemical Society 2012.

In the works introduced so far, the functional devices are based on entangled networks of nanowires. However, as explained in Section 2.1.2, many researchers in the community have seen the use of individual nanowires as a path to make the best use of 1D nanostructures in functional devices. For this reason, several attempts have been made to align, orient, and separate nanowires. Dielectrophoresis has been already mentioned as one of such methods [80]. Liu *et al.* introduce many more alignment techniques – including microfluidics, surface pre-patterning, self-assembly, and templating – and summarize their advantages and disadvantages [82]. Most of these methods are not mature enough for fully deterministic placement of individual nanowires at required locations, but still take the community closer to the fabrication of few-nanowire or single-nanowire devices.

One approach that is particularly effective in aligning and separating nanowires is microfluidics [81]. Flowing a stream of fluid containing GaP, InP, or silicon nanowires across a substrate's surface allows a highly oriented deposition of nanowires in the direction of the flow. The degree of alignment, as indicated by the angular spread of nanowires in degrees, and nanowire density are adjustable by manipulating the flow rate and flow duration, respectively. With their highest flow rate of 10 mm/s, the authors demonstrate an angular spread of 6°. In addition, by keeping the flow duration below 30 minutes, they also obtain an average nanowire-nanowire separation of approximately 400 nm (Figure 2.8c [81, 82]). A repetition of this method with different flow directions can create even complex junctions of nanowire arrays. For example, a two-step process involving orthogonal flow directions can create perpendicular nanowire arrays. However, the scalability of this method is constrained by the microchannel size, which makes uniform alignment in larger channels challenging.

Carbon nanotubes

Four main approaches for non-deterministic integration of carbon nanotubes are as follows:

1. Solution-based assembly (Figure 2.9a): Carbon nanotubes are dispersed in a solution and assembled onto pre-patterned electrodes on a device substrate. To improve the alignment of carbon nanotubes with the electrodes, techniques like dielectrophoresis [87] and selective surface modification [83] can be applied. For instance, Park *et al.* demonstrate high-density, selective, and aligned carbon nanotube integration on dielectric substrates using surface-selective modifications [83]. However, these modifications or surfactants used to disperse and align the nanotubes are hard to eliminate completely after integration, often leaving residues that could unpredictably affect the surface. This residue issue is particularly problematic for applications in sensing or quantum technology, in which clean nanotube surface is desirable.

2. Floating catalyst chemical vapor deposition (FC-CVD) (Figure 2.9b): In this method, individual carbon nanotubes are synthesized under precise conditions within a FC-CVD reactor right above the device substrate. Freshly grown nanotubes are transported by a carrier gas to the outlet of the reactor and directly "sprayed" onto pre-patterned bottom contacts as they exit the reactor. This technique allows the device substrate to remain nearly at room temperature, ensuring a chemical-free, room-temperature integration of carbon nanotubes into devices and compatibility with various substrate types. Following nanotube deposition, substrates can be thermally annealed to enhance nanotube-metal contact quality. Wei *et al.*

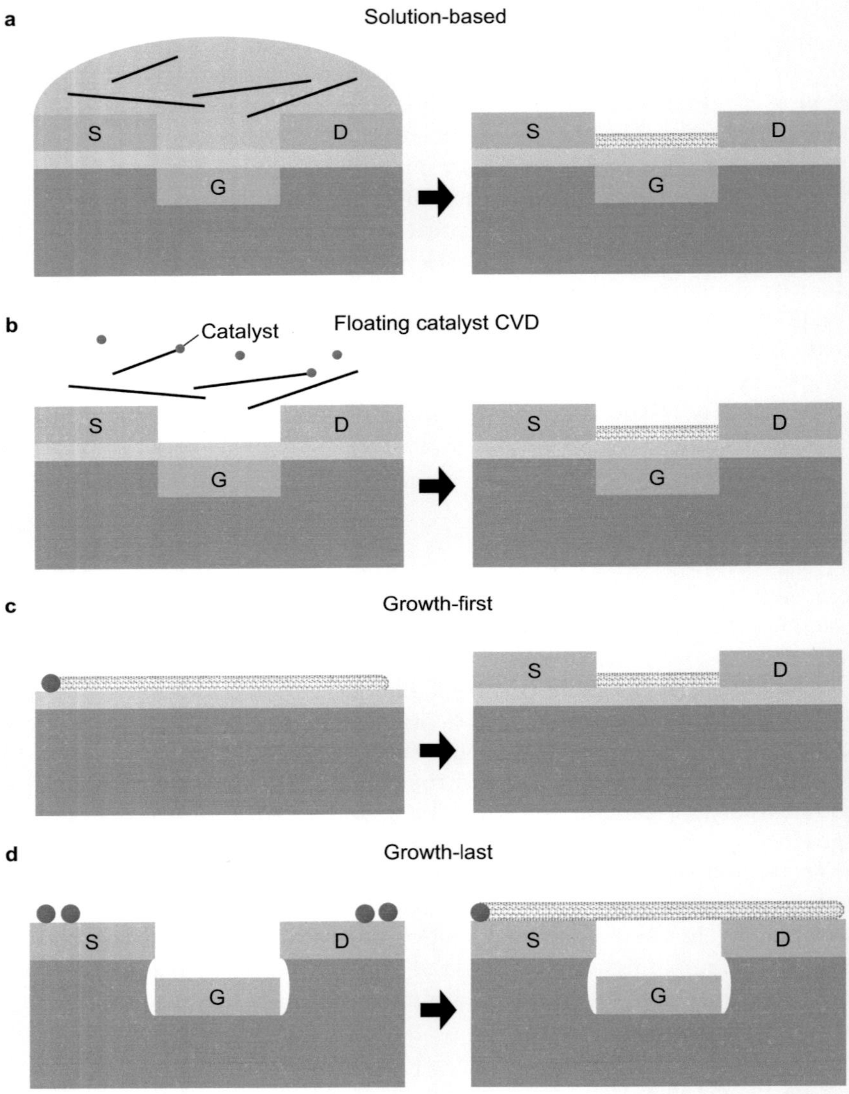

Figure 2.9: Non-deterministic integration of carbon nanotubes. (a) Solution-based assembly [83]. (b) Floating catalyst CVD (FC-CVD) [84]. (c) Growth-first integration [85,86]. (d) Growth-last integration [27,28].

demonstrate parallel fabrication of 188 CNTFETs within 3 hours, on 1000 pairs of pre-patterned Pd contacts [84]. Among these 188 devices, 144 are semiconducting devices and 44 are metallic devices. Their Monte Carlo simulation shows that, when the contact array design and nanotube synthesis conditions are tuned correctly, approximately 90 % of fabricated FETs can be single-CNT devices. However, this simulation is not verified empirically. Furthermore, the device resistances of fabricated FETs are rather high — with the median being 1.6 MΩ and the interquartile range being 1.9 MΩ – despite 2 hours of thermal annealing after nanotube integration.

3. Growth-first integration (Figure 2.9c): Nanotubes are synthesized directly on device substrates using localized catalyst areas, then top metal contacts are added through photolithography [85, 86]. The process is straightforward and requires only one or two iterations of lithography, especially if doped silicon/SiO_2 substrate can work as a back gate. This method leaves the nanotube channel open to the environment – which is desirable for applications like gas sensing – and allows certain, yet not very strong, control over nanotube orientation via flow or lattice alignment on crystalline substrates like quartz or sapphire. However, it does not allow nanotube selection, resulting in variations in nanotube length, density, and behaviors. In addition, as is the case for solution-based assembly, the formation of top contacts by lithography contaminates nanotube surfaces.

4. Growth-last integration (Figure 2.9d): Nanotubes grow directly on pre-patterned bottom contacts in the last step of device fabrication. This way, nanotubes preserve their ultraclean, pristine state. In addition, unlike the three methods listed above, this method can also make devices with individual, *suspended* nanotubes, the advantages of which are explained in Section 2.1.3. The separation of the nanotube channel from the substrate reduces gate hysteresis and electronic noise significantly, and this is especially desirable for chemical sensors and quantom dots [27]. However, the metal contacts may interfere with the CVD growth of nanotube and are exposed to the high temperatures of the CVD growth process (>800 °C). This can severely limit the choice of contact metal. As a workaround, Tormo-Queralt *et al.* [27] and Rechnitz *et al.* [28] use a "fast-heating" CVD technique, in which the device substrate with Pt contacts is quickly inserted into the heated area of the CVD furnace after the target growth temperature (900 °C) is reached. Then, the device is taken out only after 6-20 minutes, to minimize the exposure of Pt contacts to the high temperature. They demonstrate double and triple quantum dots in suspended nanotubes grown this way, but they report a very low (2 %) yield of devices with single suspended nanotube, due to the random nature of CVD-based nanotube growth. The authors suggest that they obtain, on average, one high-quality suspended CNT devices per day.

Table 2.3 presents a qualitative summary of the 4 methods introduced above. Quantitative comparison in terms of device yield and throughput will be

Table 2.3: Qualitative comparison of non-deterministic carbon nanotube integration methods.

	Solution-based assembly	**Floating catalyst CVD**	**Growth-first**	**Growth-last**
Pre-selection	No	No	No	No
Ultraclean	No	Yes	No	Yes
Suspended device	No	No	No	Yes

made in Section 2.2.3, along with deterministic nanomaterial integration methods.

2D materials

Like 1D nanostructures, 2D materials can also be dispersed in a solution and then assembled onto a device substrates [88]. The dispersion is prepared by liquid-phase exfoliation (LPE), which breaks 2D materials into 2D nanosheets in a solution via extended sonication. This enables the direct transfer of 2D nanosheets onto integrated chips through methods like drop casting, spin and spray coating, offering a quick and straightforward way to form 2D material-based films at large scale. This method, however, often results in thicker layers and has limitations in achieving film uniformity and producing films thinner than 100 nm. To enhance uniformity, multiple solution dropping cycles may be necessary.

2.2.2 Deterministic integration

As described earlier, for 1D nanostructures, the term *deterministic* corresponds to the placement of an individual nanowire or nanotube on a target spot. For 2D materials, it means the placement of a monolayer material on a target spot.

Nanowires

Individual silicon nanowires can be fabricated by conventional top-down micromachining techniques. For example, Section 2.1.4 already introduced a H_2 sensor with an individual, suspended silicon nanowire made on a silicon-on-insulator (SOI) wafer [69]. However, MOX nanowires need a post-growth device integration step.

In Section 2.2.1, dielectrophoresis, or electric field-assisted assembly, was introduced as a method that aligns nanowires to a certain extent, but not perfectly. It is true that dielectrophoresis is, in most cases, not a deterministic assembly method for individual nanowires, but when the density of a nanowire suspension is lowered significantly, it can result in an impressively high yield of single-nanowire devices. The method applies an electric field to align nanowires by using the disparity in dielectric constants between the nanowires and their medium, which is usually a liquid. Freer *et al.* [89] use this technique to consistently position single nanowires between electrodes, with a 98.5 % success rate across more than 16,000 sites on a 400 mm^2 area. The work also introduces a way to quantify assembly in electric fields, enhancing dielectrophoretic assembly models and showing its viability for precise, efficient nanomanufacturing.

Besides a few exceptions including [89], deterministic assembly of individual nanowires usually relies on a mechanical approach like pick-and-place using a micromanipulator [90–92]. For example, in [92], an individual GaP nanowire is mechanically picked up and placed between two suspended membranes for a fundamental ballistic phonon transport study. GaP nanowires are synthesized using the VLS method with gold catalyst droplets. Electron beam lithography is used to obtain nanowires with diameters below 40 nm. They suggest that increasing the V/III element ratio fivefold during nucleation and twofold during growth–compared to the typical V/III element ratio used for larger nanowires–is the key to obtaining such small-diameter nanowires. After synthesis, an individual nanowire are positioned across membranes using a micromanipulator and investigated with Raman thermometry.

,A recent study by Ali *et al.* reports a breakthrough in the precision and universality of the pick-and-place assembly [93]. They utilize polyethylene terephthalate (PET) monofilaments, which are finely tapered from 200 μm to sub-micron thickness using a custom heat-and-pull method (Figure 2.10). The filament is brought close to a hot surface at 240 °C for melting. Subsequent pulling of the filament away from the hot surface leads to the cooling of the liquid polymer, which forms a sharp taper at the tip of the filament. This process is precisely controlled by adjusting the pull length and velocity on a motorized stage. The tapered tip, the diameter of which is as small as 0.38 μm, is then mounted on a micromanipulator to work as a nanoscale crane that picks up an individual nanowire from a vertical forest. When tip comes very close to an individual nanowire, the nanotube "jumps" into contact with the PET tip due to van der Waals interactions. As long as the tip diameter is below 2 μm, this approach consistently picks up single InP, AlGaAs, and GeTe nanowires with diameters between 120 nm and 400 nm with around 80 % success rate. The authors suggest that the minimum nanowire diameter is partly limited by the optical microscopy resolution. The picked-up nanowire stamped or placed on a device substrate with pre-patterned bottom contacts. The nanowire can also be suspended between the contacts, resulting in an individual, suspended nanowire device, the advantages of which have already been outlined in Section 2.1.4.

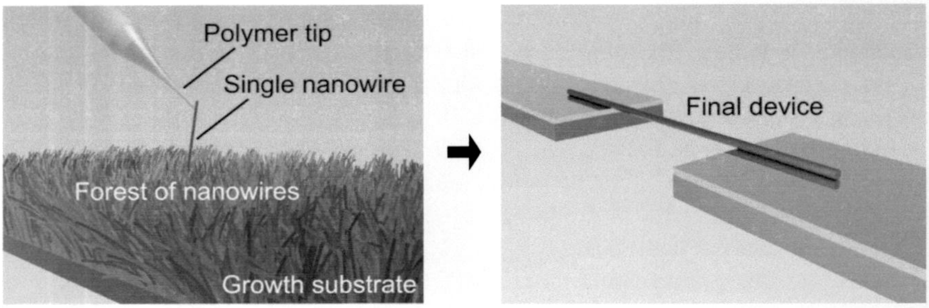

Figure 2.10: A tapered polymer filament tip picks up an individual nanowire from a forest of vertical nanowires and places it on a device substrate with pre-patterned bottom contacts [93]. Reproduced with permission from [93]. © Wiley-VCH GmbH 2022.

Carbon nanotubes

Even the tapered polymer filament tip presented by Ali *et al.* [93] is too thick to pick up an individual carbon nanotube, which is >100 times smaller than nanowires. However, a similar idea of picking a nanotube with a sharp tip has been demonstrated with atomic force microscopy (AFM) tips, which have the radius of curvature smaller than 10 nm [94]. However, production of high-quality pick-up tips is difficult, and the pick-up process is also very time-consuming. As a result, this approach has not been adopted widely.

Instead, deterministic integration of individual carbon nanotubes has been mostly realized by a "grow-and-place" approach (Figure 2.11) [29]. This process starts with the growth of individual, suspended nanotubes between cantilevers on a growth substrate (Figure 2.11a). Then, the cantilever pairs can optionally be scanned by scanning electron microscopy (SEM) [95] or Raman spectroscopy [29] to locate suspended nanotubes. Raman spectra of a carbon nanotube can be used to pre-select nanotubes that are desirable for the intended device application based on metallicity, diameter, and defectiveness [96] (Figure 2.11b). Finally, selected nanotubes are mechanically or robotically transferred onto pre-patterned electrodes on a device substrate, without any exposure to processing chemicals (Figure 2.11c). The suspended, ultraclean nanotubes assembled in this or a similar way are particularly interesting for applications such as gas sensors [25, 60], nanomechanical resonators [30, 97–99], and quantum dots [100–102].

The robotic assembly process (Figure 2.11c) involves three key steps. Initially, the process begins with the accurate lateral alignment of a carbon nanotube – or the cantilever gap that hosts the carbon nanotube – with the target device structure. If needed, angular alignment can be added to this step. Subsequently, the growth substrate approaches the device for the transfer of the carbon nanotube. This transfer step is also called *stamping* in some studies. For serial fabrication of multiple devices, this procedure is iteratively repeated. So far, all studies in the literature use manually controlled micromanipulators equipped with piezo stages. As a result, this robotic assembly remains a time- and labor-intensive process with limited device throughput (<1 device/hour).

However, one can imagine a fully *automated* operation of these three steps. A computer-controlled robotic system fed with the coordinates of carbon nanotubes on the growth substrate and the coordinates of target devices can automatically (i) align Nanotube 1 and Device 1, (ii) transfer the nanotube onto the device, and (iii) iterate these steps for the next (Nanotube, Device) pairs with pre-saved coordinates. If made possible, such automated robotic assembly can significantly improve the throughput and reduce the manual workload of the process.

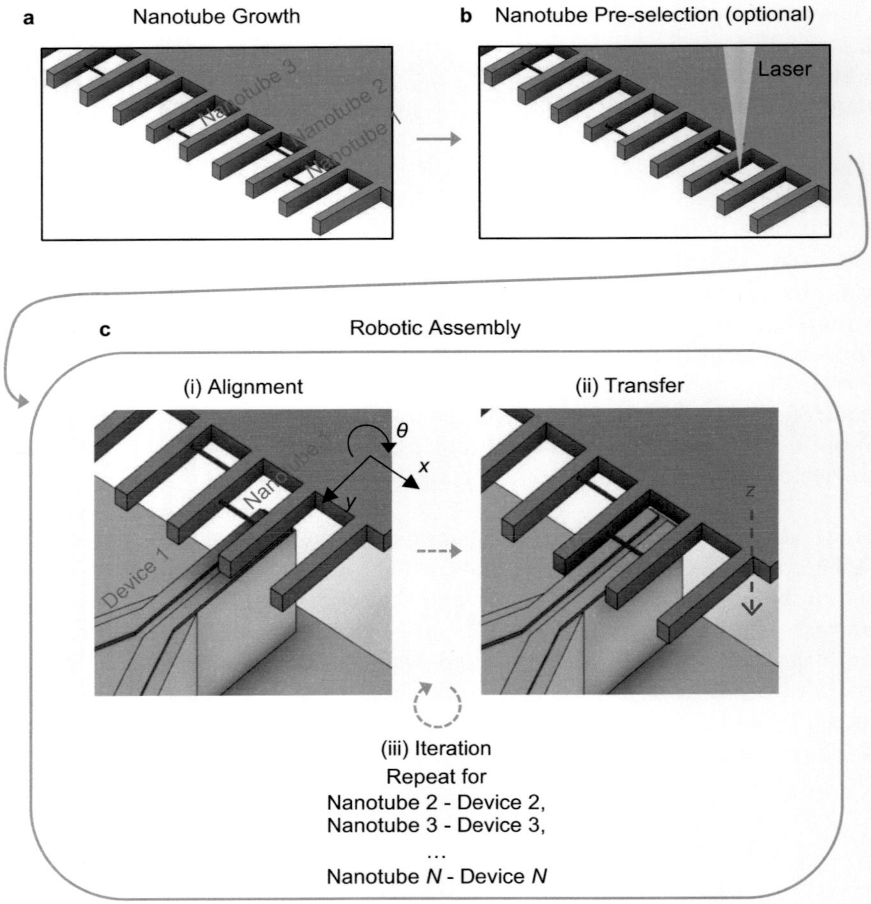

Figure 2.11: Grow-and-place assembly of individual carbon nanotubes. (a) Growth of individual, suspended carbon nanotubes on a dedicated growth substrate. (b) Optional pre-selection of carbon nanotubes with Raman spectroscopy. (c) Mechanical transfer of a nanotube onto a pre-patterned device. An automated robotic system can be used to repeat this transfer process to assemble multiple devices in a serial manner.

2D materials

In the 2D material research community, mechanical transfer or stamping is used much more widely for deterministic material integration. It also involves the three steps illustrated in Figure 2.11c – alignment, transfer, and iteration – but a notable difference is that 2D material flake is attached to the surface of a transfer stamp or cantilever, instead of being suspended between cantilevers like carbon nanotubes.

A vast number of papers report different stamp materials and pick-up techniques, and a comprehensive review of these topics can be found in [88, 103]. Here, attention is paid to robotic assembly setups or systems, which are most often used to stack layers of 2D materials van der Waals superlattices or heterostructures. As is the case for carbon nanotube assembly setups, most of the 2D material transfer systems – including commercial systems offered by HQ Graphene Systems BV and CreativeDevices – are not automated yet.

However, a couple of recent papers start to report partially or fully automated 2D material transfer systems. Masubuchi *et al.* demonstrate a robotic system that automatically determines the position of an exfoliated 2D flakes and assembles them into superlattices inside a glovebox [104]. The assembly cycle starts after a user defines the superlattice design. In each alignment step, the selected 2D crystal – either graphene or hBN – is brought into the target spot with a lateral error of 10 μm. Human intervention is needed between the alignment step and the transfer step, as the operator has to manually confirm and, if needed, fine-tune the alignment. The authors report a throughput of 4 stacking iterations per hour, with a few minutes of human intervention needed for each iteration. Although not fully automated, this system marks an important advance toward automated robotic assembly of nanomaterials.

Mannix *et al.* take a step further and demonstrate a fully automated assembly system, which transfers 20-25 pixelated layers of 2D materials per hour in high vacuum [105]. Their vacuum assembly robot, which they call VAR in short, is equipped with a heated x-y translation stage that accommodates multiple source substrates and receiving substrates of about 1 cm^2. Additionally, it includes a z-θ actuated platform for stacking, employing a microstructured polymer stamp to lift and stack the 2D materials. By operating inside a compact, desktop-sized vacuum chamber, which maintains a pressure around 10^{-6} torr), the VAR keeps surface contamination and oxidation to a minimum and ensures high interfacial quality between 2D material layers. An integrated optical microscope and digital camera enable precise lateral alignment and real-time detection of contact between the stamp and substrate. The versatility of this assembly robot is demonstrated with transfers of MoS_2, WS_2, WSe_2, and hBN, and the full automation of the system is highlighted by an overnight, operator-free fabrication of a 80-layer MoS_2 solid stacked layer-by-layer.

2.2.3 Summary: robotic assembly of individual nanomaterials

Discussions in Section 2.2.2 make it clear that robotic assembly is adopted more and more widely as a deterministic integration method for individual carbon nanotubes and 2D materials. So far, this approach has been considered manual, slow, and therefore not suitable for high-volume device fabrication, but recent advances in the automation of 2D material robotic assembly systems contribute to noticeable improvement in assembly throughput.

Section 2.2.2 already proposed the idea of a fully automated robotic assembly system for carbon nanotubes. Chapter 4 of this thesis describes the system in further detail. The development of the system is not within the scope of this thesis. Instead, this thesis focuses on analyzing the assembly result of the automated robotic assembly system and characterizing the CNTFETs assembled by the system. The automated robotic assembly system described in this thesis is compared in Table 2.4 with other best-in-class robotic systems used for the assembly of individual carbon nanotubes and monolayer 2D materials (see "This work" in table). The table includes two non-deterministic, parallel integration methods for individual carbon nanotubes for reference. The comparison suggests that full automation can improve the throughput of robotic assembly to a level that is comparable to a non-deterministic, parallel integration method (FC-CVD [84]), while obtaining a significantly higher yield.

2.3 Objectives of this work

The main objectives of this thesis are as follows:

- **Improve the yield of low-resistance CNT devices.** As described in 1.2.1, improving CNT-electrode contact quality is a key challenge in this direction.
- **Characterize devices fabricated by a high-throughput process flow based on automated robotic assembly of carbon nanotubes.** Demonstration of suspended CNT devices fabricated by automated robotic assembly of carbon nanotubes and the characterization of device properties.
- **Demonstrate fast readout and recovery of carbon nanotube**

gas sensors at ultra-low power. Demonstration of an ultra-low-power sensor that can complete its recovery-readout cycle within minutes.

The importance of these objectives was highlighted in Section 2.1.6 and 2.2.3, in which the gaps in the research of nanostructure-based gas sensors and nanomaterial integration methods were discussed.

The subsequent chapters address the objectives of the thesis as follows.

Chapter 3 introduces an effective contact improvement method for bottom-contacted carbon nanotube devices. The proposed method involves

Table 2.4: Robotic assembly of individual nanomaterials. For comparison, two non-deterministic, parallel integration methods (marked with asterisks) are also presented. Yield is defined differently for 2D materials and CNTs. For 2D materials, yield indicates the success rate of robotic transfers, as verified by optical microscopy. For CNTs, yield indicates the ratio of electrically active devices, as defined by the presence of a measurable source-drain current (I_{sd}) without a short to the gate. N/A: Not available.

Work	Material	Automation	Throughput (/hour)	Yield
Masubuchi *et al.* [104]	2D materials	Partial automation (alignment only)	4 transfers	~100 %
Mannix *et al.* [105]	2D materials	Full automation	25 transfers	~100 %
Waissman *et al.* [102]	Suspended CNT	No automation	<1 transfer	N/A
This work	Suspended CNT	Full automation	46 transfers	70 %
*Wei *et al.* [84] (FC-CVD)	Substrate-bound CNT	N/A	60 devices	18 %
*Tormo-Queralt *et al.* [27] (Growth-last)	Suspended CNT	N/A	<1 device	2 %

a two-step process: pre-contacting cleaning of bottom electrodes using Ar-ion and post-contacting thermal annealing. After a brief literature review of various contact improvement approaches, the experimental procedures for this work's proposed method are detailed. The obtained results are analyzed and compared with results of other studies in the literature. In addition, the correlation between the contact improvement and the self-heating operation of carbon nanotube-based gas sensors is discussed.

Chapter 4 covers the automated robotic assembly of ultraclean carbon nanotubes into devices. The discussion includes the conceptualization of the new fabrication process, tool sets and substrate designs, and characterization of the automated transfer itself and of the devices fabricated by the new process.

Chapter 5 reports fast and ultra-low-power operation of suspended carbon nanotube-NO_2 sensors. The proposed operation method enabled by nanotube self-heating and transient sensing. The self-heating effect in the nanotube channel significantly accelerates the desorption of gas molecules, reducing the sensor recovery time to a minute. As gas molecules re-adsorb on the nanotube after cooling, the initial rate of the sensor transient is used to determine NO_2 concentration within a few minutes. This work discusses temperature profiles along the self-heated carbon nanotube, their effect on different sensing regions, and a physical model-based fit.

The final chapter (Chapter 6) summarizes the conclusions of this work. The summary includes not only the main achievements, but also a critical discussion about the limitations of this work. Finally, it proposes future directions of research.

3 Contact Improvement for Nanotube Self-heating

In Section 1.2.1, the poor electrical contact between the nanotube and the electrodes has been pointed out as one of the technical challenges that hinder the application of CNT-based NO_2 sensors. This chapter introduces an effective contact improvement method for bottom-contacted carbon nanotube devices. The proposed method involves a two-step process: pre-contacting cleaning of bottom electrodes using Ar-ion and post-contacting thermal annealing. After a brief literature review of various contact improvement approaches, the experimental procedures for this work's proposed method are detailed. The obtained results are analyzed and compared with results of other studies in the literature. In addition, the correlation between the contact improvement and the self-heating operation of carbon nanotube-based gas sensors is discussed. A significant portion of the content in this chapter is reported in [106], and some of the work presented in this chapter is an outcome of collaborations with Roland Hauert, Miroslav Haluska, and Cosmin Roman.

A color version of this chapter is available at https://doi.org/10.3929/ethz-b-000702474.

3.1 Carbon nanotube-electrode contact

The use of carbon nanotubes (CNTs) as channels in field-effect transistors (FETs) has turned carbon nanotube's exceptional material properties into several practical devices [107–109]. Despite these advances, the substantial contact resistance ($R_{contact}$) at the interface between the CNT and the electrode remains as one of the most commonly discussed problems that limit the overall performance of CNT devices [110].

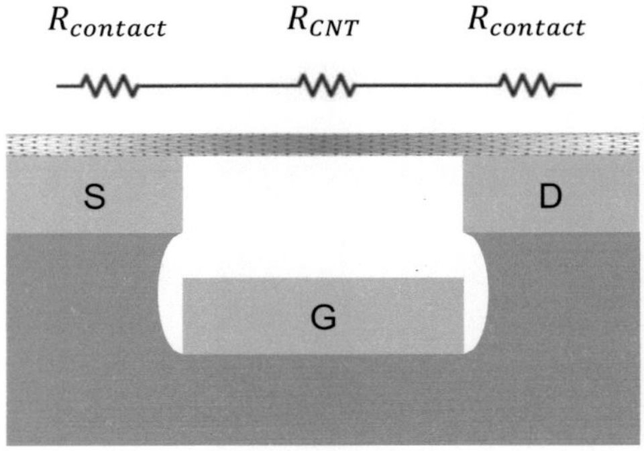

$$R_{device} = 2R_{contact} + R_{CNT} + R_Q$$

Figure 3.1: Schematic illustration of the overall device resistance (R_{device}) of a carbon nanotube field-effect transistor (CNTFET), which can be understood as a series of quantum resistance (R_Q), channel resistance (R_{CNT}), and two contact resistances ($2R_{contact}$).

As illustrated in Figure 3.1, the total resistance (R_{device}) of a carbon nanotube field-effect transistor (CNTFET) can be simply modeled as a series of quantum resistance (R_Q), channel resistance (R_{CNT}), and two contact resistances ($2R_{contact}$). R_Q, which corresponds to the quantum limit of resistance in a carbon nanotube, is known to be ~6.45 kΩ [111, 112]. R_{CNT} of nanotubes grown by chemical vapor deposition (CVD) typically ranges in the tens of kΩ per μm of nanotube [113, 114]. However, the R_{device} values reported in the literature often exceed the combined contribution of R_Q and R_{CNT} by a significant margin [83, 84, 115], often showing R_{device} above 1 MΩ for devices with few-μm nanotube channels. Such wide distribution of R_{device} strongly suggests that the predominant limiting factor in R_{device} is the contact resistance ($R_{contact}$).

3.1.1 Contact in bottom-contacted devices

Based on the contacting methods, CNTFETs can be categorized into two groups: top-contacted and bottom-contacted devices. The former is fabricated by placing a nanotube on a substrate and subsequently depositing metal electrodes onto it, and studies in the transistor community mostly use such top-contacted, substrate-bound FETs [116, 117]

(Figure 2.5 left). On the other hand, the fabrication of ultraclean devices for gas sensors [24,25], electromechanical resonators [97,118], and quantum devices [95, 119] is predominantly realized with bottom-contacted devices, especially the ones, in which an individual nanotube is suspended between source and drain contacts. Mostly, these devices are fabricated by growing individual, suspended CNTs on a growth substrate and then mechanically transferring them onto pre-patterned source and drain electrodes on a device substrate [29] (Figure 2.5 right).

While the literature continuously witnesses active discussions about ideal contact materials [120, 121], electronic structure [122], and contact improvement techniques [123], many of these discussions are based on substrate-bound, top-contacted devices, as they are the mainstream architecture in the transistor community. However, the challenges in bottom-contacted devices require different considerations. Establishing an effective nanotube-electrode contact in this configuration is not trivial, as the surface conditions of the pre-patterned bottom electrodes depend on various factors including processing and storage history, which add complexity to the contacting process. Several contact improvement methods for bottom-contacted devices have been proposed, including electron beam-induced deposition (EBID) of graphitic carbon interconnect [124] and electron beam soldering [125]. They appear to be effective in reducing the contact resistance (R_{CNT}), these electron beam-based techniques may introduce amorphous carbon and potentially contaminate the otherwise clean surface of the nanotube. Additional top metallization through selective atomic layer deposition (ALD) has been explored to enhance bottom contacts, but its effectiveness is sometimes limited, especially for devices with poor initial contact [115].

All of these methods focus on improving the contact quality after contacting, leaving gaps in our understanding of whether and how the nanotube-electrode contact can be improved by pre-contacting treatments. In this context, the method presented here offers new insights into the need for effective pre-contacting treatments.

3.1.2 Low contact resistance for nanotube self-heating

Section 2.1.4 suggested that the self-heating, or Joule heating, of 1D nanostructures is a interesting phenomenon for the low-power operation of 1D nanostructure-based gas sensors. This phenomenon becomes particularly pronounced in the case of suspended CNTFETs, in which minimal heat dissipation into the substrate leads to substantial self-heating at just a few microwatts, which is indicated by negative differential conductance (NDC) in the output characteristic (I_d-V_{sd}) of a suspended nanotube [61].

The ability of suspended CNTs to undergo significant heating at very low power paves the way for ultra-low-power recovery of CNT-based gas sensors. Chikkadi *et al.* reports complete desorption of NO_2 from a self-heated, suspended nanotube in 10 minutes at 3 μW [25].

However, the effectiveness of self-heating is critically dependent on the contact resistance. When the contact resistance is high, excessive voltage drop at contacts can possibly prevent the suspended nanotube channel from heating up. The investigation in this chapter later supports this claim empirically by showing that the NDC is observed only in devices with resistance below a certain value.

3.2 Device fabrication and contact improvement methods

This section describes the fabrication process of the suspended CNTFETs used in this chapter. Figure 3.2 provides a graphical overview of the steps, which are detailed in the subsequent subsections.

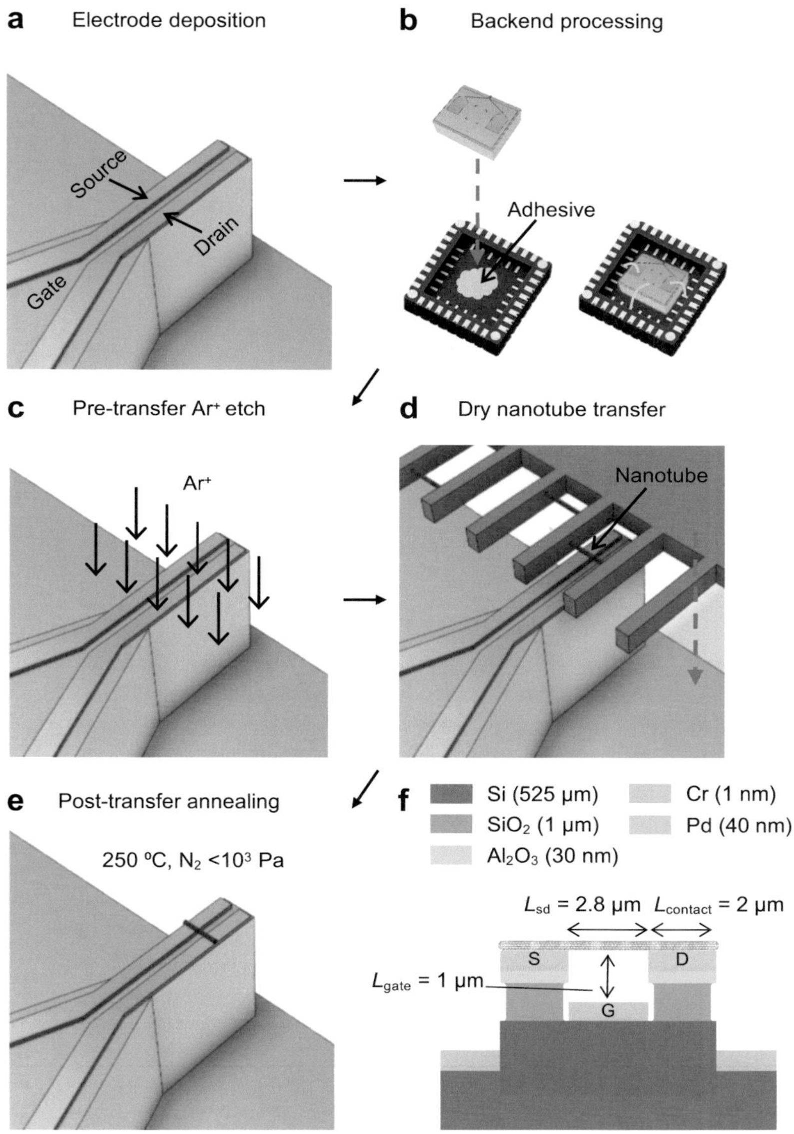

Figure 3.2: (a) Illustration of the device substrate. Source, drain, and gate electrodes are deposited by e-beam evaporation. (b) The backend processing involves die bonding with an adhesive and wire bonding. (c) Pre-transfer electrode surface cleaning with Ar-ion sputter etch. (d) Room-temperature, dry transfer of suspended nanotubes onto the electrodes. (e) Post-tranfer thermal annealing. (f) Schematic illustration of suspended CNTFETs used in this chapter. Legend indicates layer thicknesses. Reproduced from the author's own publication [106] (open access available under the Creative Commons CC-BY-NC-ND license).

The fabrication of suspended CNTFETs in this work primarily employs the "grow-and-place" method (Section 2.2.2) to integrate suspended carbon nanotubes onto the device substrate [29] and applies two contact improvement measures before and after the dry transfer. The subsequent subsections will introduce the two substrates required for this technique – growth substrate and device substrate – and how the suspended nanotubes are transferred from the growth substrate to the device substrate.

3.2.1 Growth substrate

Fork-like growth structures are fabricated by photolithography standard surface micromachining with a silicon-on-insulator (SOI) wafer. The device layer is 5 μm thick, and the surface is covered with $\sim$100 nm wet thermal oxide. A fabrication run sheet for the growth substrate can be found in Appendix A.

As shown in Figure 3.3, the growth substrate consists of 16 *forks* on its circumference [126]. This design, described in a patent application [126], distinguishes itself from existing growth substrate designs in the literature, which have a fork only on one side [95, 100, 127]. Having forks on multiple sides of the growth substrate is effective in scaling the number of suspended CNTs grown on the substrate. In addition, when the CNTs grown on different forks are dry-transferred onto devices for device integration, the growth substrate can simply be rotated. Each fork hosts several cantilevers, which are also referred to as *fingers* in this thesis. Individual, suspended nanotubes are grown between the fingers, and this space between the fingers is also referred to as a *trench* in this thesis. There are approximately 560 trenches on each growth substrate.

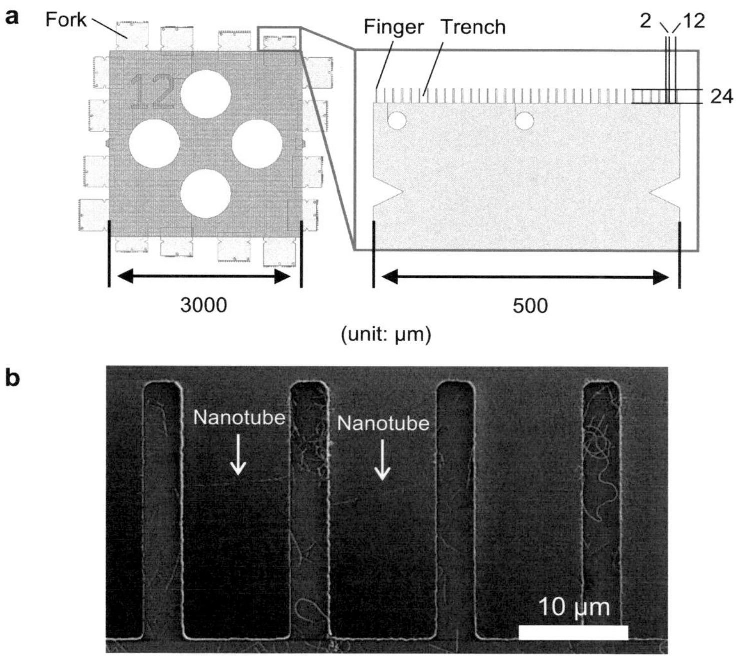

Figure 3.3: (a) Graphical illustration of a growth substrate. The substrate consists of 16 forks on its circumference, and each fork hosts a few tens of cantilevers, also referred to as fingers [126]. (b) Scanning electron microscopy (SEM) image of a fork-like growth substrate with suspended nanotubes grown between cantilevers. Reproduced from the author's own publication [106] (open access available under the Creative Commons CC-BY-NC-ND license).

3.2.2 Growth of suspended CNTs

The nanotubes are grown through catalytic chemical vapor deposition (CVD). For this, iron-loaded ferritin particles, which are size-separated by sedimentation velocity centrifugation at 40000 RPM for several hours [128], are deposited onto the growth substrate [129]. This deposition is performed by simply immersing the growth substrate in a diluted ferritin solution for 4 minutes. After the immersion, the growth substrate is rinsed with deionized water for 1 minute. To oxizide the protein shells in ferritin particles and obtain iron oxide (Fe_2O_3), the growth substrates are heated on a ceramic plate by a Bunsen burner for 10 minutes. Then, iron catalyst particles are obtained by reducing Fe_2O_3 with H_2 at 825 °C for 2 minutes. This reduction step is performed in the CVD oven directly before the nanotube growth step. Finally, the nanotubes are grown in an atmosphere consisting of CH_4 and H_2 gases at specific pressures (330 mbar for CH_4 and 220 mbar for H_2), with a temperature maintained at 825 °C for 30 minutes. These steps are illustrated in Figure 3.4. The CVD oven used in this study is Aixtron BM Pro 6, the schematic illustration of which is available in Figure 3.5.

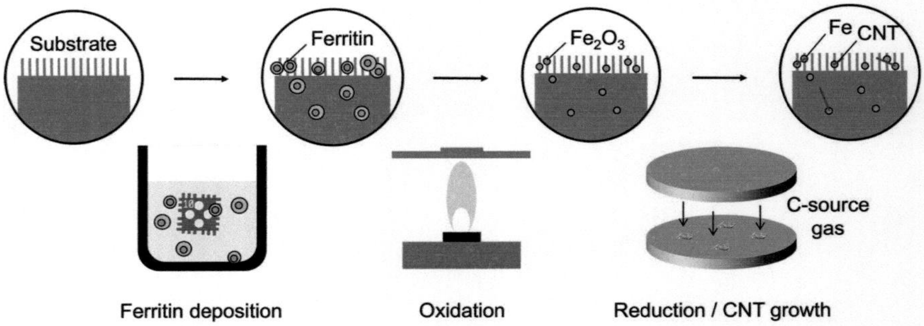

Figure 3.4: Illustration of the process steps for catalytic CVD growth of suspended carbon nanotubes. This is a re-created drawing of an original illustration from [129].

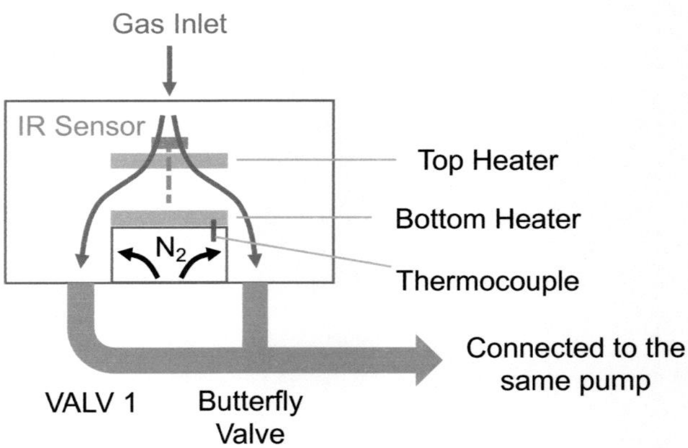

Figure 3.5: Schematic illustration of Aixtron BM Pro 6, a CVD oven used in this work.

3.2.3 Device substrate

The fabrication of the device substrate starts with a 4-inch silicon wafer, on which 1-μm wet thermal SiO_2 is grown, followed by a deposition of Al_2O_3 via atomic layer deposition (ALD) at 300 °C. The mesa structures and electrodes are patterned using inductively coupled plasma (ICP) etching. This step is followed by vapor HF etching to create an undercut [130]. The source, drain, and gate electrodes are then deposited using e-beam metal evaporation (1 nm Cr and 40 nm Pd).

All devices reported in this chapter have the following geometrical dimensions: L_{sd} (source-drain pitch) = 2.8 μm, $L_{contact}$ (contact width) = 2.0 μm, and L_{gate} (gate-CNT distance) = 1.0 μm (Figure 3.2f).

The completed device substrate is attached to a ceramic-gold package using a silver paste (Acheson Silver DAG 1415), which is cured at 180 °C (453 K) for 10 minutes. Electrical connections are then established with Al wire bonds. The full process sheet for the device substrate can be found in Appendix A.

3.2.4 Pre-transfer electrode etching with Ar-ion

Most of the studies in the literature focus on improving the electrode-nanotube contact quality after contacting, but this work focuses on cleaning the electrode surface before contacting. This is done by physically etching the electrode surface with Ar-ion with a sputter tool (Von Ardenne CS 320 S) at 25 W for 60 seconds, at a pressure below 1 Pa. The etch rate of the tool was calibrated for SiO_2 (1.01 nm/min), and this corresponds to an approximate etch rate of 2.3 nm/min for Pd, based on the sputter rate ratios reported in the literature [131].

3.2.5 Dry transfer of CNTs

Individual suspended nanotubes, grown as previously described in 3.2.2, are transferred onto the source and drain electrodes of the device. This transfer is facilitated by a piezo-driven micromanipulator (Figure 3.6) and monitored through in-situ current measurements with Keithley 2400 source measurement units. An SEM image of a suspended CNTFET is presented in Figure 3.7.

For the devices reported in this chapter, the alignment of the growth substrate and device substrate and the transfer motion (Figure 3.2c) are performed manually by a user controlling the piezo stages. This technique is referred to as *manual transfer* throughout this thesis, as opposed to *automated robotic transfer* that is described in Chapter 4.

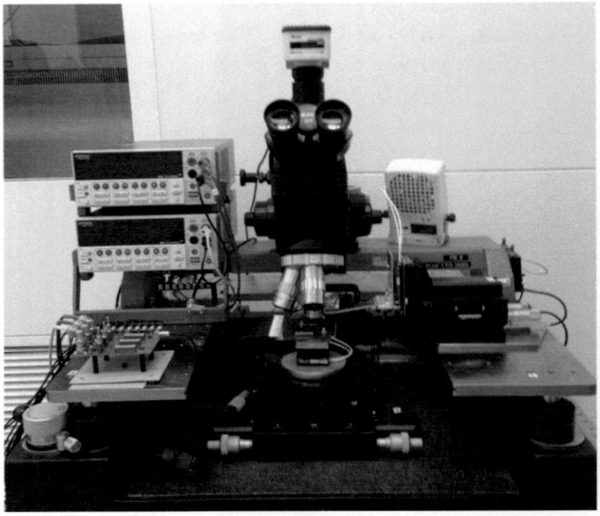

Figure 3.6: Photograph of the piezo-driven manual micromanipulator containing micrometer screws, piezo-stages, optical microscope, and electrical readout components used for in-situ current monitoring. Image courtesy of Sebastian Eberle.

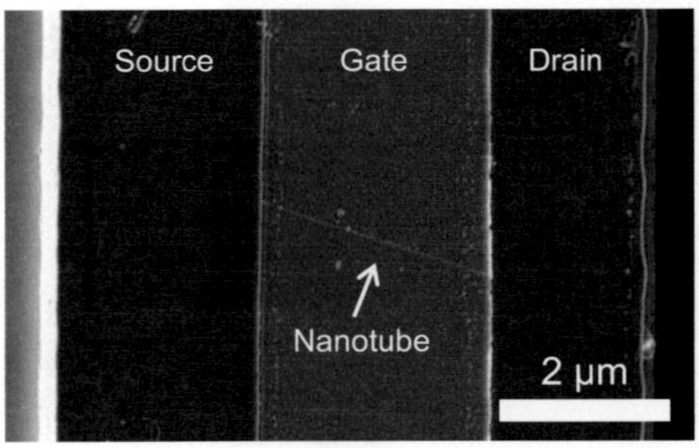

Figure 3.7: SEM image of a suspended CNTFET fabricated by dry transfer. Reproduced from the author's own publication [106] (open access available under the Creative Commons CC-BY-NC-ND license).

3.2.6 Post-transfer thermal annealing

Post-contacting thermal annealing is known to improve the electrode-nanotube contact due to reduced potential barrier and desorption of adsorbates from the interface region [132–134]. Therefore, we combine this method with the previously described pre-transfer electrode cleaning (Section 3.2.4) to further optimize the electrode-nanotube contact. Thermal annealing in this work is performed in a N_2 atmosphere at 250 °C below 10^3 Pa for 1 hour (Figure 3.2e).

3.3 Results and analysis

3.3.1 Contact improvement results

To investigate the effect of pre-transfer electrode etching with Ar-ion, three independent device groups are compared. Group 1 is the control group of devices without the contact improvement methods described in 3.2.4 and 3.2.6. Group 2 devices are treated with pre-transfer electrode etching with Ar-ion, but not with post-transfer thermal annealing. Group 3 devices are treated with both contact improvement methods.

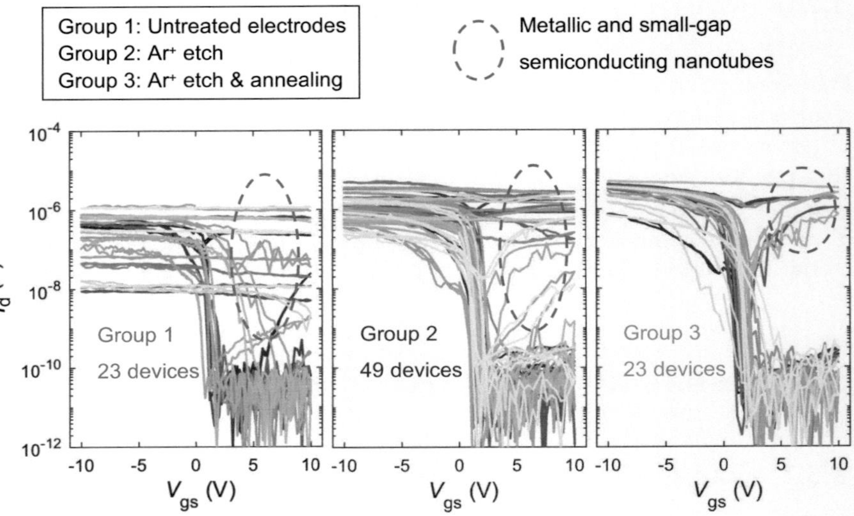

Figure 3.8: Transfer characteristics of all devices measured at V_{sd} = 0.5 V. The definitions of Group 1, 2, and 3 follow the descriptions in the main text. Reproduced from the author's own publication [106] (open access available under the Creative Commons CC-BY-NC-ND license).

Figure 3.8 shows the transfer characteristics (I_d-V_{gs}) of all devices in the three groups. For Group 1, a median ON-resistance (R_{on}) measured at V_{gs} = -10 V and V_{sd} = 0.5 V is 1.56 MΩ, with a wide interquartile range (IQR) of 9.38 MΩ. Group 2 devices, which were treated with an additional Ar-ion etching step, showed a considerable reduction in both median R_{on}, with a value of 456 kΩ, and interquartile range, with a value of 482 kΩ. It is noteworthy that such significant contact improvement was obtained without any post-transfer processes. Post-transfer thermal annealing led to further contact improvement in Group 3 devices, which showed even lower median R_{on} of 143 kΩ and a narrower interquartile range of 59 kΩ.

The transfer characteristics of devices across all three groups clearly show an overall increase and a narrowed distribution of ON-currents in Groups 2 and 3. The empirical cumulative distribution functions of the R_{on} in Figure 3.9 visualize these trends more clearly. In the literature, devices treated only with post-contacting annealing exhibit a much larger distribution of resistance that the Group 3 devices of this work, suggesting that the effect of annealing is limited if the initial contact is poor [83,84,115]. Here, due to the good initial contact established by the pre-transfer electrode cleaning, 87 % of the devices were found to have R_{on} values below 250 kΩ.

In Figure 3.9, semiconducting devices are distinguished from metallic devices. The transfer characteristics of semiconducting carbon nanotubes (CNTs) and metallic carbon nanotubes carbon nanotubes exhibit distinct behaviors due to their differing electronic properties. Semiconducting CNTs display a clear gate-dependent ON/OFF behavior ($I_{on}/I_{off} > 10^2$), in which the current (I_d) increases or decreases significantly with changes in gate voltage (V_{gs}). In contrast, metallic CNTs show little to no gate dependence, maintaining a relatively constant current regardless of the gate voltage due to the absence of a bandgap. Between the two groups lie small-gap semiconducting CNTs, which have a very narrow bandgap and, as a result, exhibit intermediate characteristics. While they do respond to gate voltage changes, their I_d-V_{gs} curves are less pronounced compared to wide-bandgap semiconducting CNTs. This small bandgap allows some current leakage even in the OFF state. For this reason, their transfer characteristics typically have a "V" shape with both n- and p-branches. In Figure 3.9, small-gap semiconducting devices are marked as metallic, as they behave more similarly to metallic CNTs in terms of R_{on}. The same categorization of CNTs is used throughout this thesis.

To understand if the R_{on} differences introduced by the contact improvement methods are statistically meaningful, two-sample Kolmogorov-Smirnov test (K-S test) with 5 % significance level has been applied to Group 1 and 2, and subsequently, Group 2 and 3. The two-sample K-S test is used to determine the likelihood that two sets of samples are drawn from the same but unknown probability distribution. This test is particularly useful for comparing the empirical distribution functions of two samples, thereby quantifying the distance between them using the K-S statistic. The test returns an asymptotic p-value, which represents the probability of observing a K-S statistic as large as the observed value under the null hypothesis that the two samples are drawn from the same distribution. The p-value takes sample size into account and becomes more accurate as sample sizes grow.

The results of the two-sample K-S tests indicate significant differences introduced by the contact improvement methods. For Group 1 and Group 2, the null hypothesis was rejected with a K-S statistic of 0.60 and a p-value of 9.10×10^{-6}. Similarly, for Group 2 and Group 3, the null hypothesis was also rejected with a high K-S statistic of 0.73 and a p-value of 3.75×10^{-8}.

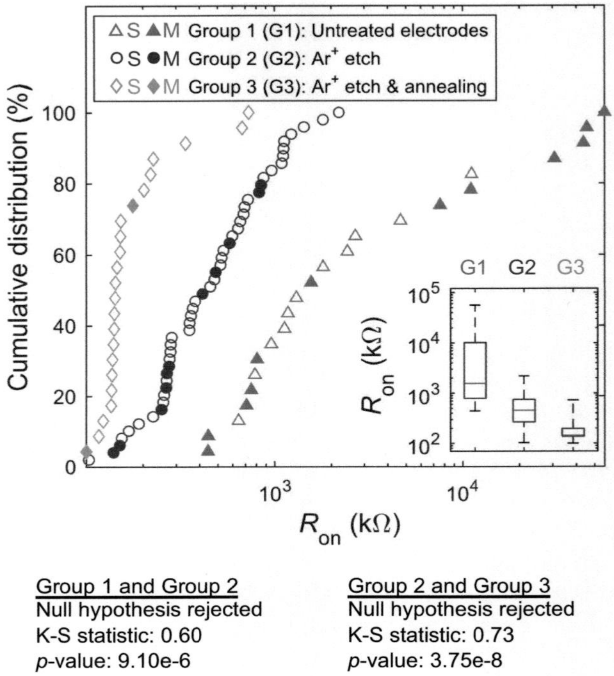

Figure 3.9: Empirical cumulative distributions of R_{on} of all devices measured at V_{gs} = -10 V and V_{sd} = 0.5 V. Empty markers represent semiconducting devices, and filled markers represent metallic devices. Inset: boxplots showing the 0th, 25th, 50th, 75th, and 100th percentiles of the distributions. The results of two-sample Kolmogorov-Smirnov tests are summarized on the bottom. Reproduced from the author's own publication [106] (open access available under the Creative Commons CC-BY-NC-ND license).

For semiconducting devices with high ON/OFF ratios, a further analysis of other performance indicators such as threshold voltage (V_{th}) and subthreshold swing (SS) is also performed.

V_{th} is defined as V_{gs} that corresponds to a device current of 2 nA. Figure 3.10a presents the empirical cumulative distribution functions of V_{th} across the three groups introduced earlier, with the median and IQR indicated for each group. The median and IQR changes as follows: V_{th} of 1.18 ± 0.81 V for Group 1, 1.31 ± 0.51 V for Group 2, and 1.62 ± 0.72 V for Group 3.

The results of the two-sample K-S tests for the V_{th} distributions indicate that the null hypothesis is not rejected for both Group 1 versus Group 2 (K-

S statistic: 0.28, p-value: 0.52) and Group 2 versus Group 3 (K-S statistic: 0.31, p-value: 0.17), suggesting no significant difference in V_{th} distributions between these groups.

Figure 3.10b illustrates the empirical cumulative distributions of SS. Group 1 shows a median SS of 180 ± 235 mV/dec (IQR), Group 2 has a median SS of 195 ± 198 mV/dec, and Group 3 has a median SS of 200 ± 103 mV/dec.

Similarly, the two-sample K-S tests for the SS distributions show that the null hypothesis is not rejected for both Group 1 versus Group 2 (K-S statistic: 0.22, p-value: 0.86) and Group 2 versus Group 3 (K-S statistic: 0.20, p-value: 0.78), indicating no significant difference in SS distributions between these groups.

The K-S tests performed on R_{on}, V_{th}, and SS suggest that the effect of the contact improvement methods on CNTFETs is largely limited to the tunneling behaviors at contacts. The three device groups share the same device architecture and geometrical dimensions. One sees that, when these factors are kept the same across device groups, V_{th} and SS distributions maintain notable similarities, even when R_{on} changes significantly.

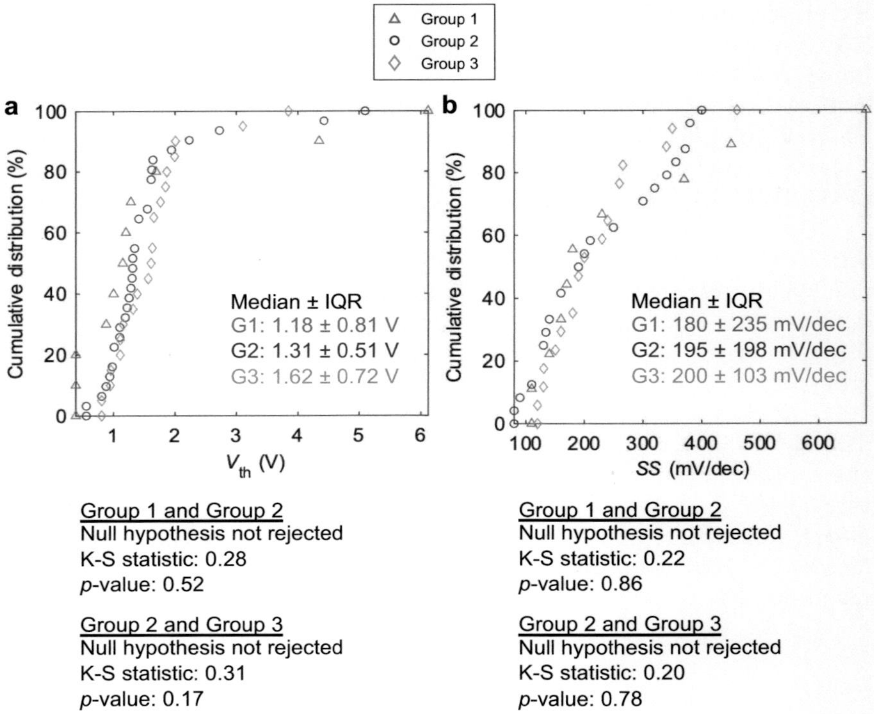

Figure 3.10: Empirical cumulative distributions of (a) threshold voltage (V_{th}) and (b) subthreshold swing (SS) of semiconducting devices. The results of two-sample Kolmogorov-Smirnov tests are summarized on the bottom.

3.3.2 Electrode surface analysis with XPS and AFM

It is important to understand why the nanotube-electrode contact is so poor in the Group 1 devices and how the Ar-ion electrode cleaning improves the contact so significantly in the Group 2 devices. Therefore, the chemical modifications of the electrode surface were assessed using X-ray photoelectron spectroscopy (XPS).

First, freshly deposited Pd electrodes (after Figure 3.2a) and those that went through backend processing steps (after Figure 3.2b) were compared. In both cases, only oxygen and carbon were detected besides Pd. The fresh Pd electrodes showed atomic concentrations of 12.8 % oxygen (as determined by O KLL) and 35.2 % carbon (as determined by C 1s). After backend processes, the O KLL concentration went up to 20.9 % and the

C 1s concentration decreased relatively to 22.9 %. To validate whether the changes are mainly coming from the use of silver paste or the heating at ~450 K, another device heated at 450 K for 10 minutes without the silver paste. The atomic concentrations were determined to be similar – 21.4 % for O KLL and 24.1 % – to those observed after all the backend processing steps. This suggests the observed changes in the atomic concentrations were predominantly due to the heating of the sample at 450 K, not due to the use of silver paste.

With a consideration that the chemical changes introduced by the sample heating could be a cause of the poor nanotube-electrode contact in Group 1 devices, the next experiment aimed to understand how the Ar-ion sputter etching changes the chemical composition of electrode surface. This was done by a depth profile analysis with the X-ray spectrometer, which alternates between Ar-ion sputter etching and and spectra acquisition. Each etch step corresponds to removal of 3.4 nm Pd, according to the calibrated etch rate of the instrument (7.39 nm/min for Ta_2O_5) and the ratios of sputter rates of different materials reported in the literature [131, 135]. After the etching of the first 3.4 nm, most oxygen and carbon were eliminated, indicating that the backend processing-induced surface changes were present mostly on the very top surface of the electrodes and they could be reversed with brief Ar-ion sputtering. In other words, a short Ar-ion etch effectively resets the electrodes to almost 100 % Pd.

Next, it is interesting to understand how the atomic concentrations evolve over time after the surface resetting via Ar-ion etching and to examine the potential correlation between this evolution and nanotube-electrode contact quality. First, the *baseline* atomic concentrations, which originate from the spontaneous adsorption of carbon and oxygen, were investigated. The sample was cleaned with Ar-ion, and then transferred to the XPS spectrometer's fast-entry loadlock, into which 99.5 % purity N_2 was introduced for one minute. The loadlock was then evacuated again to achieve high vacuum conditions over the course of an hour. Following this, XPS spectra from the sample, referred to as the '0 min' sample in Figure 3.11a and Figure 3.12a-c, were obtained. The oxygen concentration after this short exposure to an N_2-filled environment was 5.3 %, and the carbon concentration was 28.8 %. This methodology was adopted from Greczynski *et al.* [136].

In the next set of experiments, the sample was cleaned with Ar-ion, exposed to controlled laboratory air (295 K, 40 % R.H.) for different time periods, and then measured by XPS. These procedures were repeated several times with varying exposure time (0.5-12 hours) to laboratory air. After starting at 5.3 %, the oxygen content slowly increases over time, reaching 20.9 % after approximately 12 hours. The carbon content remains around 30 % (Figure 3.11a). The changes of core-level spectra, as illustrated in (Figure 3.12a-c), reveal that the changes in the overall oxygen content can be attributed to four chemical changes occurring at different rates: (1) spontaneous adsorption of carbon compounds (initiated within a few seconds), (2) adsorption of oxygen-containing species (a few minutes), (3) oxidation of adsorbed carbon or adsorption of species containing a C-O bond (1-12 hours), and (4) oxidation of Pd (1-12 hours). In the first hour, an increase in overall oxygen content is observed, but signs of C-O bonds (Figure 3.12c) or Pd oxidation (Figure 3.12a) are not present. These two chemical changes are only observed in the spectra obtained after a 12-hour exposure to air.

Similar experiments were repeated with shorter exposure times (10-30 minutes) at higher temperatures (450 K and 600 K). The heating was done by placing the sample on a hot plate. At these temperatures, increase in oxygen content is observed much faster. An oxygen content of 21.4 % is already reached after 10 minutes at 450 K, surpassing the value observed after a 12-hour air exposure at 295 K (Figure 3.11a). The quick rise in oxygen content can be attributed to the accelerated adsorption of oxygen-containing species, as C-O bonds and the oxidation of Pd are not yet evident (Figure 3.12a, c). These changes are also not observed after 20 and 30 minutes of heating. At 600 K, the changes are more drastic; oxygen content increases to 31.5 %, 34.7 %, and 36.9 % after 10, 20, and 30 minutes of heating, respectively. Here, the oxidation of Pd is clearly visible in Figure 3.12a. As oxygen predominates on the surface, the relative carbon content is found to be lower on the heated samples than on the sample stored at room temperature.

These experiments can be used to interpret the previously described results in Figure 3.8. Firstly, the rapid increase in oxygen content at higher temperatures is consistent with what was observed after the backend process that includes a 10-minute curing of an adhesive (silver paste) at 450 K. Here, the major contribution to the high oxygen concentration is the adsorption of oxygen-containing species, rather than surface oxidation. Secondly, oxygen is adsorbed onto the electrode surface slowly at room temperature after oxygen and carbon are removed by a short Ar-ion etch, while the surface carbon concentration almost immediately returns to its pre-etching value. Therefore, the contact improvement effect of the electrode etching in Group 2 devices is attributed to the removal of oxygen, rather than that of carbon. In Figure 3.11c, the ON-resistances of Group 2 devices are depicted as a function of time between Ar-ion etching and nanotube transfer (i.e. time between Figure 3.2c and Figure 3.2d), and the devices are grouped into three time windows for comparison (Figure 3.11c inset). The changes in the 25th, 50th, and 75th percentiles indicate that the effect of electrode etching diminishes slowly over time. This can also be explained by the slow increase in oxygen content on the electrode surface at room temperature (Figure 3.11a).

Oxygen-containing species are expected to change local surface energy and reduce the metal's work function when adsorbed on a polycrystalline Pd surface [137]. The formation of an interlayer between Pd and CNT can also be expected. Moreover, the presence of oxygen at the CNT-metal interface is known to result in the formation of a surface inversion layer and, consequently, a locally higher tunneling barrier [138]. These chemical changes help to understand how Ar-ion etching removes the source of poor nanotube-electrode contacts.

However, the physical effect of Ar-ion sputtering on the electrode surface should also be considered. Atomic force microscopy (AFM) scans performed with Bruker FastScan in tapping mode on untreated (Group 1) and Ar-ion-treated (Group 2) bond pads show that Ar-ion etching results in an increase in surface roughness (as indicated by arithmetic average of roughness profile, R_a, which is the average of the absolute values of surface height deviations from the mean line) of Pd electrodes from 0.56 nm to 0.72 nm (Figure 3.13). Note that the AFM scans were performed on bond pads due to the small size of the electrode area, which is only 2 μm wide. Larger surface roughness is considered to hinder a nanotube's ability to conform to the local topography of a substrate [137], so the physical effect of Ar-ion sputtering is actually adverse to nanotube-electrode contacts. This result suggests that the chemical effect of resetting the electrode surface is strong enough to counteract the unfavorable roughness change induced by the Ar-ion etch.

All XPS measurements described so far were conducted with the Quantum 2000 spectrometer from Physical Electronics equipped with an monochromatic Al-Kα X-ray source (hν = 1486.7 eV) and a hemispherical electron energy analyzer set at a 45° electron escape angle relative to the surface plane. The focus area diameter was 50 μm, targeting the 200 μm-by-200 μm bond pads, not the electrodes that are only 2 μm wide, due to size constraints. Background pressure during tests was below $5{\cdot}10^{-7}$ Pa. Calibration of the electron energy scale utilized the Au $4f_{7/2}$ signal set at 84.0 ± 0.1 eV. Electron and ion neutralizers were used to reduce possible sample charging because the metallic electrodes were deposited on silicon and insulating silicon dioxide and were therefore not electrically connected to the sample platen. Compensation for any remaining surface charging was done by shifting the Pd $3d_{5/2}$ and Au $4f_{7/2}$ peaks to 335.2 eV and 84.0 eV, respectively. Elemental survey scans were taken at 117.4 eV analyzer pass energy, whereas detailed spectra of O 1s, C 1s, Pd 3d, Au 4f core levels and the O KLL Auger were acquired at 93.9 eV. These analyzer pass energies yield a total analyzer energy resolution of 1.4 eV determined for Au 4f electronics. For Pd electrodes, O 1s (529-532 eV), the most commonly used region for oxygen, could not be used due to its interference with the Pd $3p_{3/2}$ (binding energy 533 eV) region. Instead, O KLL peak was analyzed. Depth profiling was performed by alternating between a spectra acquisition and electrode surface etch with 2 kV Ar-ion across a 2 mm-by-2 mm area. The etch rate of the Ar-ion was calibrated to be 7.39 nm on Ta_2O_5. Atomic concentrations were derived from peak areas with Shirley background correction using MultiPak 8.2 software. The XPS measurements were performed by Roland Hauert, whose extensive support is sincerely appreciated.

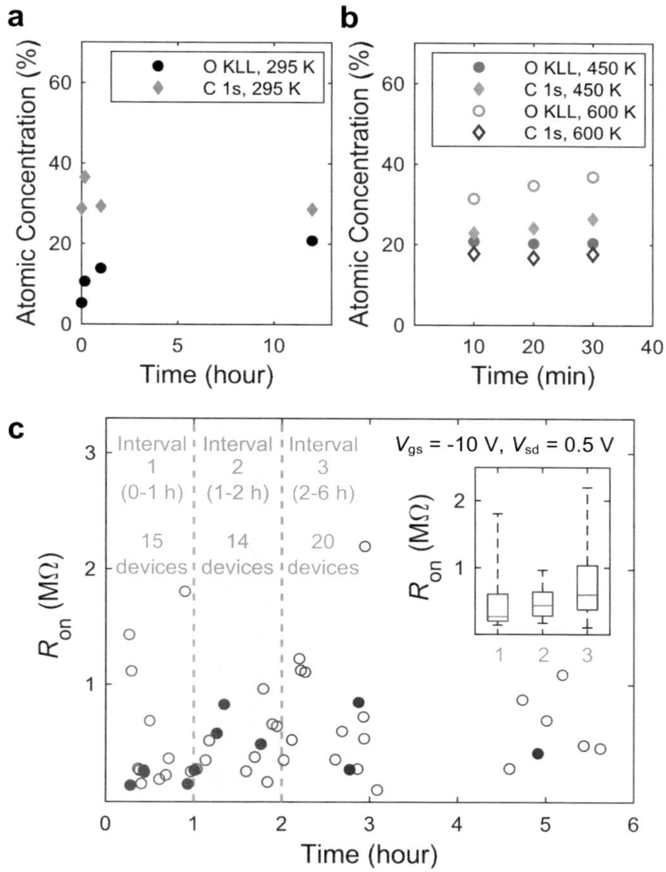

Figure 3.11: (a) Changes in surface oxygen and carbon concentrations at room temperature (295 K). (b) Changes in surface oxygen and carbon concentrations at elevated temperature (450 K, 600 K). (c) R_{on} of Group 2 devices versus time between Ar-ion electrode etching and nanotube transfer. The devices are categorized into three time windows: Interval 1 (0-1 h), Interval 2 (1-2 h), and Interval 3 (2-6 h). As in Figure 3.9, empty markers represent semiconducting devices, and filled markers represent metallic devices. Inset: boxplots showing the 0th, 25th, 50th, 75th, and 100th percentiles of the three time intervals. The increase of the 25th, 50th, and 75th percentiles over time can be correlated with the slow increase in oxygen content on electrode surface at room temperature shown in (a). Reproduced from the author's own publication [106] (open access available under the Creative Commons CC-BY-NC-ND license).

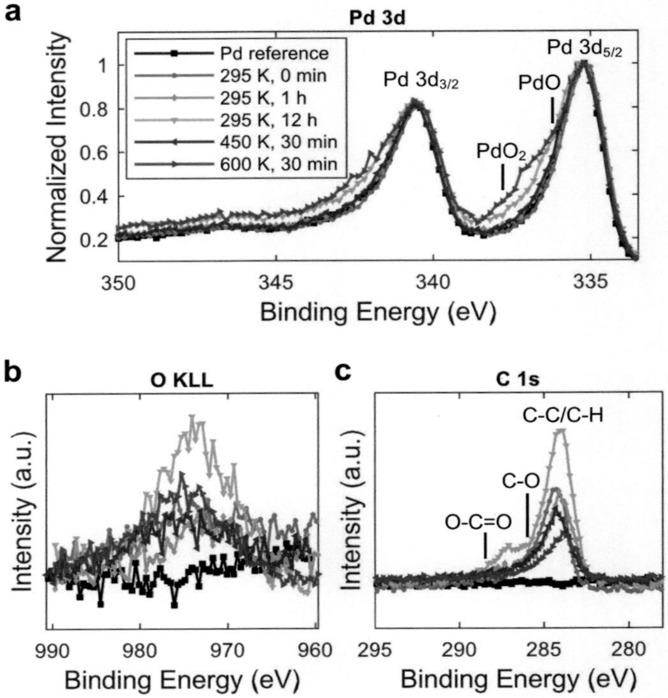

Figure 3.12: XPS core-level spectra for (a) Pd 3d, (b) O KLL, and (c) C 1s. The Pd 3d spectra in (a) are normalized to highlight the broadening of Pd $3d_{5/2}$ peak towards PdO and PdO_2. Reproduced from the author's own publication [106] (open access available under the Creative Commons CC-BY-NC-ND license).

These investigations into electrode surface chemistry and topography via XPS and AFM provide insights into the mechanisms behind the contact improvement effect of Ar-ion electrode etching.

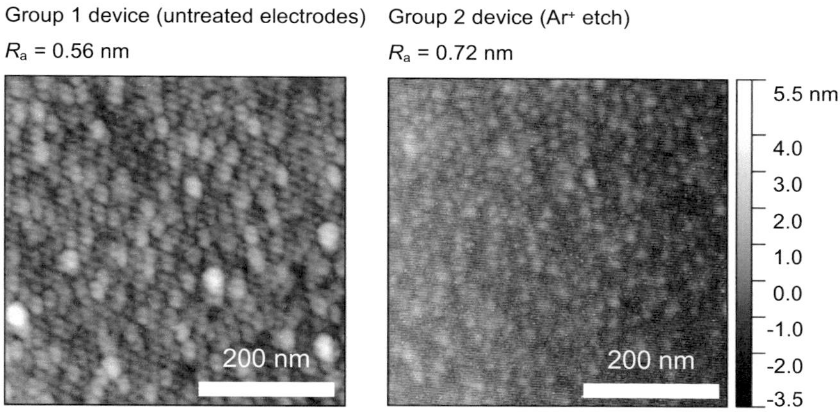

Figure 3.13: AFM images of an untreated Pd electrode (Group 1 device) and an Ar-ion-etched electrode (Group 2 device). Reproduced from the author's own publication [106] (open access available under the Creative Commons CC-BY-NC-ND license).

3.3.3 Contact quality analysis with Raman spectroscopy

Raman spectroscopy was performed to characterize and compare the quality of nanotube-electrode contacts in a bad-contact, high-resistance device (from Group 1) and a good-contact, low-resistance device (from Group 3). In the case of a Group 1 device (Device ID: SJDS83-D3, R_{on} = 1.5 MΩ), the characteristic G mode of CNTs (~1590 cm^{-1}) was detected on all parts of the nanotube, including contact sites on both electrodes (Figure 3.14a-b). There was a clear difference in G-mode intensity between the suspended part and contact-supported part of the nanotube – with an intensity ratio of 46 (suspended):1 (source) and 42 (suspended):1 (drain) – but the G mode was clearly detectable on the contacts. Similar differences in Raman intensity between the suspended part and the supported part of a nanotube have been shown in previous research [139, 140]. The decrease in intensity on the supported part is explained by changes in the electronic structure of the supported part and, as a result, changes in the resonance conditions of the Raman effect. The intensity ratio is also related to contact quality because forces of nanotube-substrate adhesion can induce changes in the Raman spectra. Thus, a very high Raman intensity ratio between the suspended and supported parts in a bottom-contacted CNTFET can be regarded as an indication of good adhesion and, consequently, good electrical contacts between the nanotube and the electrodes. In a Group 3 device (Device ID: SJDS85-D3, R_{on} = 336 kΩ), the G mode on the contact area was suppressed so much that it was not detectable despite multiple repetitions of line scans with a measurement duration of 60 seconds per spot (Figure 3.14c-d). This absence of detectable G mode suggests stronger mechanical contact between the nanotube and electrodes, and its consistency with the low device resistance indicates the improved contact quality in Group 3 devices.

In these experiments, Raman spectra were obtained using a confocal Raman microscope CRM200 from WITec. The samples were subjected to an excitation wavelength of 532.3 nm, which was polarized perpendicularly to the long axes of the device electrodes. The laser beam was focused by an objective with 100x magnification and a numerical aperture of 0.8. Scattered light was collected in the back-scattering geometry using the same objective and was then split by a grating with 1800 or 600 grooves per millimeter before being directed to a CCD detector cooled to -52 °C. The power of the laser beam was measured by a Newport optical power meter positioned above the microscope objective and was set at a moderate value of 0.4 mW to avoid inducing a heating effect on the sensitive nanotubes. The experimental support of Miroslav Haluska with Raman spectroscopy is highly appreciated.

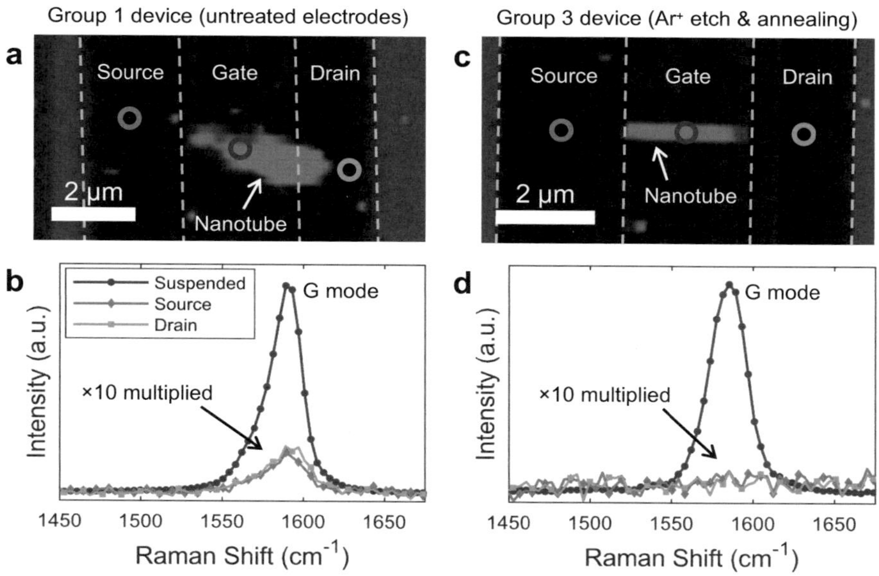

Figure 3.14: Raman spectroscopy characterization of CNT-electrode contacts. (a) A Raman image of a Group 1 device (Device ID: SJDS83-D3). Light gray is representative of the CNT's G mode. (b) Raman spectra from a suspended CNT channel and contacts of the Group 1 device. The approximate locations of the Raman scans are indicated in (a) with circles. The intensity of the spectra from the contacts has been multiplied by 10 to clearly display the G mode. (c) A Raman image of a Group 3 device (Device ID: SJDS85-D3). (d) Raman spectra from a suspended CNT channel and contacts of the Group 3 device. The approximate locations of the Raman scans are indicated in (c) with circles. The intensity of the spectra from the contacts has been multiplied by 10 to clearly display the absence of G mode. Reproduced from the author's own publication [106] (open access available under the Creative Commons CC-BY-NC-ND license).

3.3.4 Implications for nanotube self-heating

The main objective of this thesis is to fabricate suspended, bottom-contacted CNTFETs and use them as gas sensors that can quickly recover from NO_2 exposure at an ultra-low power via nanotube self-heating [25]. Therefore, it is important to understand how the improved contact quality described in Section 3.3.1 contributes to the self-heating effect of suspended nanotubes.

It has been reported that substantial heating of a suspended carbon nanotube can occur at very low power, primarily due to the absence of heat dissipation into the substrate. The effects of nanotube self-heating are manifested in the form of current saturation and negative differential conductance (NDC) in the output characteristic (I_d-V_{sd}) of a suspended nanotube [61].

However, the suspended channel cannot be heated if high contact resistances cause significant heat dissipation at contacts. Figure 3.15a displays the output characteristics of a device (Device 1, Device ID: SJDS97-D1) with a relatively high R_{on} of 384 kΩ, where no sign of current saturation is shown up to a V_{sd} of 1.2 V. When a V_{sd} sweep from 0 V to 2.1 V was applied at V_{gs} = -10 V, the nanotube was destroyed at a V_{on} of ~1.9 V, before V_{sd} reached the saturation point. This breakdown of the CNT before current saturation was repeatedly observed in a group of 11 high-resistance devices, with the lowest R_{on} among these devices being 345 kΩ.

In contrast, devices with lower R_{on} (145 kΩ for Device 2 and 160 kΩ for Device 3, Figure 3.15b) are shown to exhibit current saturation and NDC at V_{sd} below 1 V and power below 2 μW (Device IDs: SJDS95-D1 for Device 2, SJDS95-D2 for Device 3) This behavior was repeatedly observed in 10 low-resistance devices, with the highest R_{on} of these devices being 240 kΩ.

These results confirm that a contact resistance $R_{contact}$ above a certain level prevents the suspended CNT channel from heating up. Although the maximum power dissipated by Device 1 (6.36 μW) right before breakdown is significantly higher than the saturation power of Devices 2 and 3, Device 1 does not exhibit the onset of the NDC regime. This suggests that, in Device 1, voltage drop at contacts is more significant than that in the nanotube channel. The results presented here are empirical indications of the importance of low $R_{contact}$ for the self-heating operation of suspended CNT gas sensors.

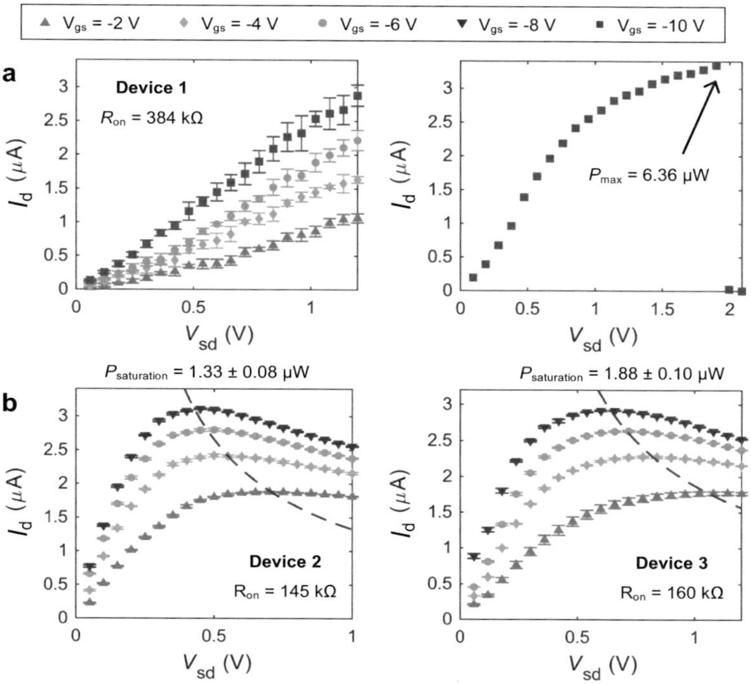

Figure 3.15: Output characteristics (I_d-V_{sd}) and nanotube self-heating. (a) Output characteristics of a device (Device 1, Device ID: SJDS97-D1) with R_{on} = 384 kΩ (left). A high V_{sd} of ~1.9 V destroys the nanotube before current saturation is observed (right). (b) Output characteristics of two devices with lower R_{on}. In both cases, current saturation is observed at V_{sd} far below 1 V, followed by negative differential conductance (NDC), indicative of significant self-heating. The current saturation points correspond to heating power P = 1.33 ± 0.08 μW (Device 2, Device ID: SJDS95-D1) and 1.88 ± 0.10 μW (Device 3, Device ID: SJDS95-D2), respectively. In all measurements, V_{sd} sweeps were repeated five times. The markers represent the mean values of the five measurements, and the error bars represent standard deviations. Reproduced from the author's own publication [106] (open access available under the Creative Commons CC-BY-NC-ND license).

3.3.5 Long-term stability

To assess the long-term stability of the nanotube-electrode contacts enhanced by pre-transfer Ar-ion etching and post-transfer thermal annealing, two Group 3 devices were kept under atmospheric conditions and their resistance changes were monitored over 201 days with electrical measurements every few weeks. The first device (Device ID: SJDS53-D2, Figure 3.16, left) exhibited a gradual yet monotonic decrease in conductance, likely due to the gradual buildup of an insulating oxide layer on the Pd surface. However, the rate of degradation decelerated, and the device maintained about 70 % of its initial conductance even after 201 days. The second device (Device ID: SJDS52-D3, Figure 3.16, right) displayed minimal degradation. Remarkably, on Day 112, its conductance was slightly higher than at the beginning. These variations in conductance are likely due to the adsorption of doping species on the channel and contacts. After Day 112, this device ceased to function and further measurements were not possible.

These findings indicate that the improved nanotube-electrode contacts in the bottom-contacted CNTFET devices demonstrate considerable stability even without contact encapsulation. Previous studies have noted that, in case of contact degradation, applying a high bias across the nanotube – a technique often referred to as current annealing – can sporadically restore device conductance through the breakdown of the oxide layer or the removal of adsorbates [19]. In this study, since the focus was on observing long-term changes under atmospheric conditions, such attempts were not made.

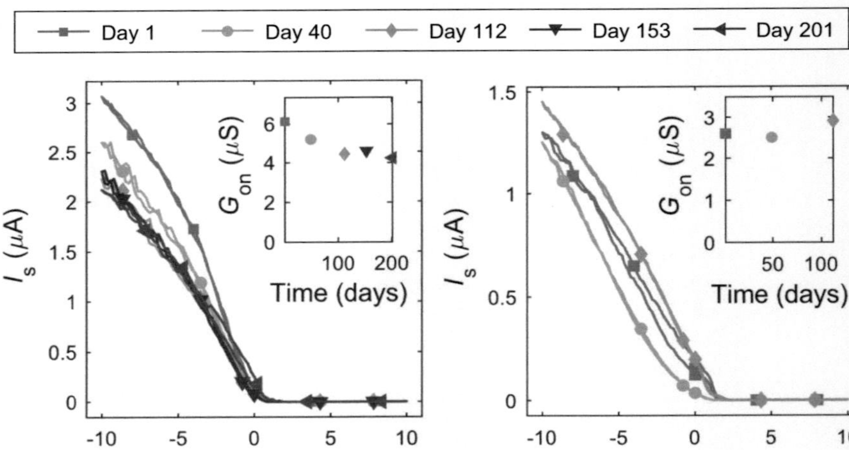

Figure 3.16: Transfer characteristics of two Group 3 devices (Device IDs: SJDS53-D2 (left), SJDS52-D3 (right)) monitored sporadically over 201 days. Insets: changes of ON-state conductance (G_{on}) measured at V_{gs} = -10 V and V_{sd} = 0.5 V. Reproduced from the author's own publication [106] (open access available under the Creative Commons CC-BY-NC-ND license).

3.3.6 Au electrodes

To test the compatibility of the contact improvement methods described so far with other types of contact metal, a similar set of experiments was performed with Au electrodes. The etching of Au electrodes with Ar-ion proved to be highly effective in lowering both the contact resistance and its variability. The median ON-resistance was decreased from 4.0 MΩ to 332 kΩ, and the interquartile range was reduced from 8.4 MΩ to 265 kΩ (Figure 3.17).

Further analysis with XPS and AFM revealed generally similar patterns for both Pd and Au electrodes except some minor differences. Similar to the findings on Pd electrodes, at room temperature (295 K), the re-adsorption of carbon-containing species on Au electrodes occurred almost immediately after surface resetting with Ar-ion, whereas the oxygen content increased gradually over 12 hours. However, surface oxygen concentrations were generally lower on Au than on Pd electrodes (Figure 3.18a). The lower oxygen content on Au electrodes could be due to the absence of oxide formation, which was not observed even after a 12-hour exposure to air. This might explain the more consistent effect of electrode etching in improving contact on Au electrodes (Figure 3.17), in contrast to the diminishing effect on Pd electrodes over after a few hours.

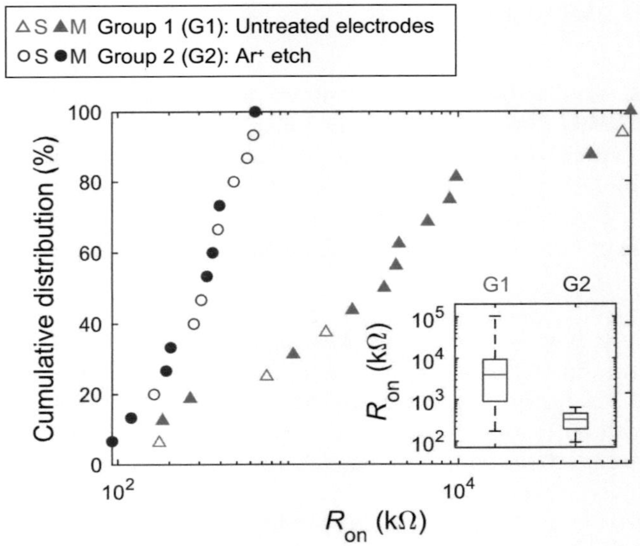

Figure 3.17: Cumulative distributions of R_{on} of all devices with Au electrodes measured at V_{gs} = -10 V and V_{sd} = 0.5 V. Empty markers represent semiconducting devices, and filled markers represent metallic devices. Inset: boxplots showing the 0th, 25th, 50th, 75th, and 100th percentiles of the distributions. Reproduced from the author's own publication [106] (open access available under the Creative Commons CC-BY-NC-ND license).

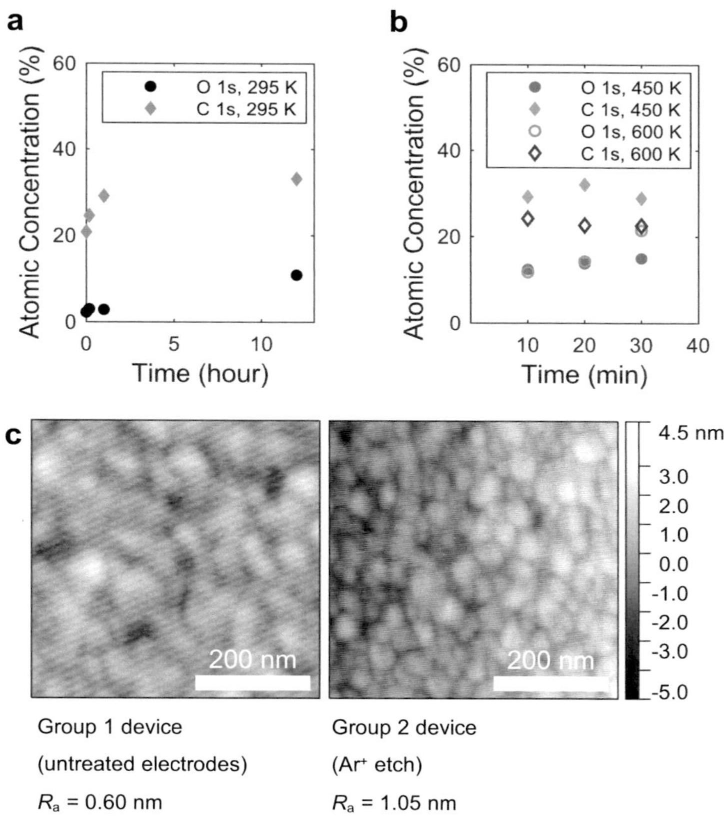

Figure 3.18: (a) Changes in oxygen and carbon concentrations at room temperature (295 K) characterized by XPS. (b) Changes in oxygen and carbon contents at 450 K and 600 K. (c) AFM images of an untreated Au electrode and an Ar-ion-etched electrode. Ar-ion etching increases R_a of Au electrodes, as is the case for Pd electrodes. Reproduced from the author's own publication [106] (open access available under the Creative Commons CC-BY-NC-ND license).

3.4 Chapter summary

In conclusion, the experiments described in this chapter suggest a significant correlation between the presence of oxygen-containing adsorbates on the

Pd or Au electrode surface and the high contact resistance at the CNT-electrode interface. Nonetheless, electrode etching with Ar-ion just prior to nanotube transfer can effectively remove these adsorbed species and reset the surface. Comparing devices without electrode treatment to those undergoing Ar-ion etching reveals that this electrode resetting improves CNT-electrode contact remarkably. This pre-transfer treatment can be further boosted by post-transfer thermal annealing. Employing both techniques together led to a reduction in the median ON-resistance of suspended CNTFETs by an order of magnitude and the interquartile range by more than two orders of magnitude. The low $R_{contact}$ achieved through these contact improvement methods proved essential for the self-heating functionality of suspended CNT gas sensors.

Table 3.1 presents a comparison of different bottom-contacted carbon nanotube field-effect transistors (CNTFETs) discussed in the literature. Unlike most prior studies that focus on post-contacting treatments, this work focuses on pre-contacting electrode cleaning, and this turned out to be the key to obtaining considerably lower R_{on} than previous work. What adds to this method's effectiveness is its consistency, which is demonstrated by the small distribution (interquartile range in the table) of R_{on}.

One interesting work to note in this table is the recent study of Althuon *et al.*, which actively adopt and refine this work's contact improvement methods and reproduce similarly low device resistances for quantum devices [100]. By cleaning the electrodes with Ar-ion before nanotube transfer and thermally annealing the devices, they see their devices reaching the open quantum dot regime, in which co-tunneling processes lead to Kondo physics phenomena and Fabry-Perot interferences at cryogenic temperatures. These phenomena are known to require extremely clean nanotube-electrode interfaces. They observe device resistance below 50 kΩ, which they consider to be sufficiently low for the open quantum dot regime, in ~7 out of 11 devices. In addition to the pre-transfer Ar-ion etching and thermal annealing proposed by this work, they introduce a few additional measures to further optimize nanotube-electrode contacts.

- Before thermal annealing, they perform preliminary current-induced annealing.
- Their radiative thermal annealing is performed at two lamp power levels: 5.6 μW and 11 μW. They observe the evolution of device resistance *in-situ*, and when the decrease of device resistance slows down at 5.6 μW, they increase the lamp power to 11 μW. This boosts a further decrease of device resistance. This experiment provides an additional insight into the correlation between heating power and contact improvement effect.

- Their process is performed in vacuum to minimize adsorbates on electrodes and the nanotube.

Along with these additional efforts, the fact that their nanotube channel is significantly shorter (~500 nm) can explain their lower device resistance values.

In general, their reproduction of clean nanotube-electrode contacts with the contact improvement methods introduced by this work emphasize that clean initial contact before CNT dry transfer is crucial for the final low device resistance.

Table 3.1: Comparison of bottom-contacted carbon nanotube devices in the literature. Median R_{on} values and interquartile ranges have been calculated or estimated from histograms and plots in each work.

Work	Woo *et al.* [133]	Jenni *et al.* [115]	Wei *et al.* [84]	Althuon *et al.* [100]	This work
Device type	L_{sd} = $1\,\mu$m, on-substrate	L_{sd} = $2\,\mu$m, suspended	L_{sd} = $2\,\mu$m, on-substrate	L_{sd} = $0.5\,\mu$m, suspended	L_{sd} = $2.8\,\mu$m, suspended
Pre-contacting treatment	-	-	-	Ar-ion electrode etching	Ar-ion electrode etching
Post-contacting treatment	Current annealing	Thermal annealing, top metallization	Thermal annealing	Current annealing, thermal annealing	Thermal annealing
No. of devices	24	7	190	11	23
R_{on} (median ± IQR)	3.8 ± 22 MΩ	0.9 ± 4.2 MΩ	1.6 ± 1.9 MΩ	47 ± 37 kΩ	143 ± 59 kΩ

4 Robotic Assembly of Carbon Nanotubes

In Section 2.2.1 and 2.2.2, various integration methods for carbon nanotubes (CNTs) were reviewed. Among the integration techniques discussed, only two methods are available for the fabrication of suspended CNTFETs, the advantages of which are described in Section 2.1.3. It was mentioned that this architecture is considered ideal for quantum devices, resonators, and, importantly for this thesis, gas sensors. The first method, direct growth, offers a straightforward approach but suffers from significant drawbacks including a very low yield [27] and high-temperature requirements that limit the choice of compatible contact metals. Notably, palladium (Pd), which is widely regarded as one of the ideal contact materials for CNTs, cannot be used in the direct growth method. The "grow-and-place" assembly described in Section 2.2.2, which is also known as *mechanical transfer* or *dry transfer*, has emerged as an alternative, offering better yield and compatibility with room-temperature processes. However, despite its advantages, mechanical transfer remains predominantly a slow and manual technique, restricting its application largely to research settings. This chapter discusses a fully automated, robotic version of this integration method. The robotic assembly is shown to overcome the limitations of manual dry transfer and demonstrates its potential as a more efficient and scalable integration process for ultraclean CNTs.

A portion of the work presented in this chapter was conducted within the framework of the Nanoassembly consortium supported by the Strategic Focus Area - Advanced Manufacturing program of ETH Board. The Nanoassembly consortium has been initiated by Prof. Christofer Hierold, whose research group had prior know-how in the dry transfer of individual carbon nanotubes [29]. The consortium aimed to develop an automated, high-throughput process flow based on this dry transfer technique, and was supervised by Prof. Michel Calame (Empa), Prof. Christofer Hierold, Dr. Cosmin Roman (ETH Zurich), Prof. Konrad Wegener (ETH Zurich), and Prof. Martin Jaggi (EPFL). Other collaborators in the consortium include Dr. Miroslav Haluska, Prof. Mickael Perrin, Dr. Jian Zhang, Rico Muff,

Dr. Rolf Brönnimann, Philipp Scheidegger, Dr. Sascha Weikert, Dr. Natanael Lanz, Andre Butzerin, and Dr. Luis Barba.

In addition to the consortium efforts, this chapter includes work that was performed in collaboration with Dr. Maria El Abbassi, Dr. Luca Ornago, Marko Nikolic, and Frederik van Veen. This chapter also introduces key results of student projects performed by Roland Bockholt [141], Julian Radziwill [142], and Kristian Ivkovic [143], which were designed and supervised by the author of this thesis.

A color version of this chapter is available at https://doi.org/10.3929/ethz-b-000702474.

4.1 Process overview

Figure 4.1 illustrates the overall process flow of the above-mentioned Nanoassembly consortium and describes the objective for each unit process.

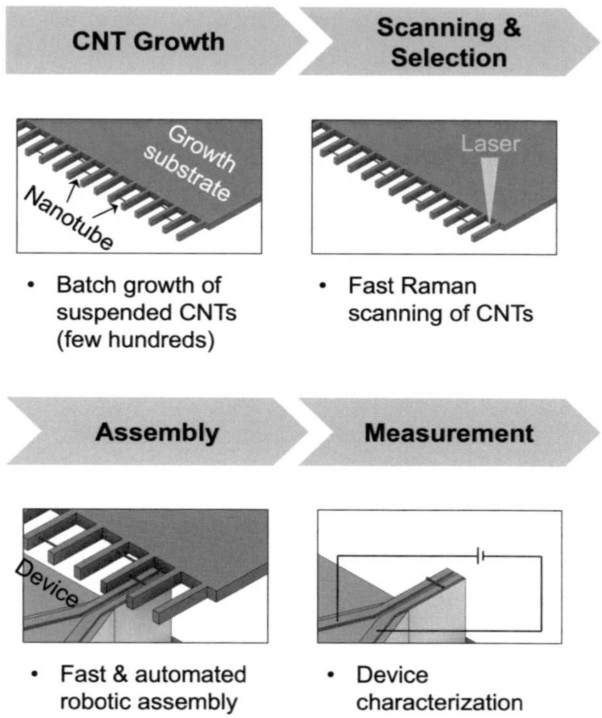

Figure 4.1: Proposed process flow for high-volume fabrication of ultraclean, suspended CNTFETs.

- **Step 1: CNT growth.** The process flow begins with the CVD growth of a large number of suspended CNTs in batch.

- **Step 2: Scanning and selection.** The CNTs are localized and selected by Raman spectroscopy. This allows *selective* assembly of carbon nanotubes in terms of e.g. number or type of carbon nanotubes, and this selective nature distinguishes this process flow from other integration techniques introduced earlier. Traditionally, Raman scanning has been considered a slow process, with each scan taking several seconds to achieve spectra with a sufficiently high signal-to-noise ratio (SNR) for reliable human identification. However, a recent work from Zhang *et al.* utilizes a convolutional neural network to considerably speed up this identification process, accurately detecting very noisy (SNR as low as 0.9) Raman spectra obtained with an integration time of 10 ms [144].

- **Step 3: Robotic assembly.** Once the CNTs are selected, they are transferred to pre-patterned device substrates using a fully automated nanoassembly machine. This machine, equipped with precision engineering and machine vision capabilities, transfers 40-50 CNTs per hour in a serial manner, without exposing the CNTs to processing chemicals.

- **Step 4: Device measurement.** The assembled devices are measured electrically. It is particularly interesting to characterize the devices' performance parameters such as R_{on}, as was done with manually assembled devices in Chapter 3 (Figure 3.9).

The scope of this thesis mainly covers the batch growth of the suspended CNTs (Step 1) and the analysis of the electrical measurement data (Step 4). The parts covered by other consortium partners will be discussed briefly and relevant references will be provided to the reader interested in technical details.

4.2 Batch growth of suspended carbon nanotubes

High-volume fabrication of suspended CNTFETs with the process flow proposed above requires batch growth of a few hundred or more suspended carbon nanotubes on a single substrate. For this objective, two key strategies were undertaken in parallel to enhance the efficiency and scale of the process. The first strategy focuses on optimizing the carbon nanotube growth parameters. The primary objective here is to maximize the chance of growing transferrable individual CNTs. This is related to the overall efficiency and potential throughput of the robotic assembly process because, if a growth substrate has a high density of transferrable individual CNTs, less of its trenches need to be skipped and the substrate needs to be replaced less often. For design terms like "trench" that are specific to the growth substrate, the reader is referred to Figure 3.3. The second strategy involves the design and fabrication of a large-scale growth substrate hosting several thousands of cantilevers. This approach is intended to substantially increase the hosting capacity of each growth substrate, which is, again, related to the throughput of the robotic assembly process.

4.2.1 Literature review

The CVD growth of carbon nanotubes (CNTs) – of all forms, e.g. substrate-bound, suspended, forests, films, etc – is a complex process significantly influenced by several parameters including reaction temperature, pressure, and gas composition. Given this complexity, the adoption of a multi-parameter optimization method is crucial for enhancing the efficacy and reproducibility of CNT synthesis. Several studies have implemented such methods to refine the growth conditions for substrate-bound nanotubes [145, 146].

However, the growth of individual, suspended CNTs poses a different challenge because simply increasing the number of CNTs does not necessarily yield positive outcomes, as it often leads to a higher number of bundled nanotubes. These bundles are less desirable as they complicate the subsequent transfer processes and integration into devices. Although optimizing growth of individual, suspended CNTs is crucial for applications that rely on the "grow-and-place" assembly method, the literature is scant on quantitative information about this process. Most studies do not provide detailed metrics such as the density of transferrable, individual nanotubes. Table 4.1 lists a few papers that grow suspended nanotubes using different catalysts and substrates, and as shown in the table, Pei *et al.* is one of the very few works that comment on the number of suspended nanotubes.

Table 4.1: Growth of individual, suspended carbon nanotubes with different catalysts and substrates.

Work	**Catalyst**	**Substrate**	**% single-CNT trenches**
Gramich *et al.* [147]	Ferritin-based Fe	Polysilicon	No information
Blien *et al.* [127]	Co	Quartz	No information
Waissman *et al.* [102]	Ti/Pt	Silicon	No information
Wu *et al.* [30]	Mo/Fe	Quartz	No information
Pei *et al.* [119]	Mo/Fe	Quartz	~10 %

To the best of current understanding, no existing studies have focused specifically on the multi-parameter optimization of suspended CNT growth, indicating a significant gap in the current research and an opportunity for contributions to the field.

4.2.2 CNT growth parameter optimization

This work aims to address the above-mentioned information gap regarding the optimization of suspended carbon nanotube (CNT) growth. Part of the experimental work reported in this section was performed by Roland Bockholt, whose bachelor thesis was supervised by the author [141]. The author produced the growth substrates, the design of which is illustrated in Figure 3.3, and ferritin catalyst solutions used in the experiments, and advised on the multi-parameter design of experiments (DoE). Roland Bockholt performed CVD growth runs and SEM scans, except for the last optimization experiment (Figure 4.5) that was performed by the author. This section summarizes the key concepts, methodology, and results of the study, and for further details including all SEM images, the reader is referred to Roland Bockholt's report, which can be provided by the author upon request.

The primary goal of this study is to maximize the number of transferrable, individual CNTs, a term that requires a clear definition within this context. The growth substrate consists of trenches that are 12 μm in width and 24 μm in length. A hypothetical 5 μm by 20 μm box, labeled as the "transferrable area," is superimposed on microscopic images of the trenches (specifically illustrated on Trench 1 and Trench 4 in Figure 4.2). This transferrable area approximates the usable space for subsequent dry transfers of CNTs in the fabrication process. Earlier, it was mentioned that the Nanoassembly process flow uses Raman spectroscopy for CNT scanning and pre-selection (Figure 4.1). For nanotubes that are actually transferred onto devices, Raman spectroscopy is preferred as a non-destructive imaging technique. However, in this study, SEM was chosen for nanotube imaging and counting because of its speed and ease of use.

A nanotube qualifies as transferrable and individual if it satisfies the following criteria:

(1) It intersects with both the left-hand side (LHS) and right-hand side (RHS) of the transferrable area.

(2) It is located at least 5 μm away from any other CNT within the same trench, measured vertically along the center line of the trench.

According to these criteria, Trench 1 and Trench 2 in Figure 4.2 each contains one transferrable nanotube, while Trench 4 contains two.

Based on the number of transferrable nanotubes identified, the trenches can be categorized into three groups:

(1) No-CNT trench

(2) Single-CNT trench

(3) Multiple-CNT trench

The best scenario for dry transfer is to grow many transferrable, individual CNTs and have as many of them as possible in single-CNT trenches, so that the robotic assembly machine does not have to transfer nanotubes in the same trench separately.

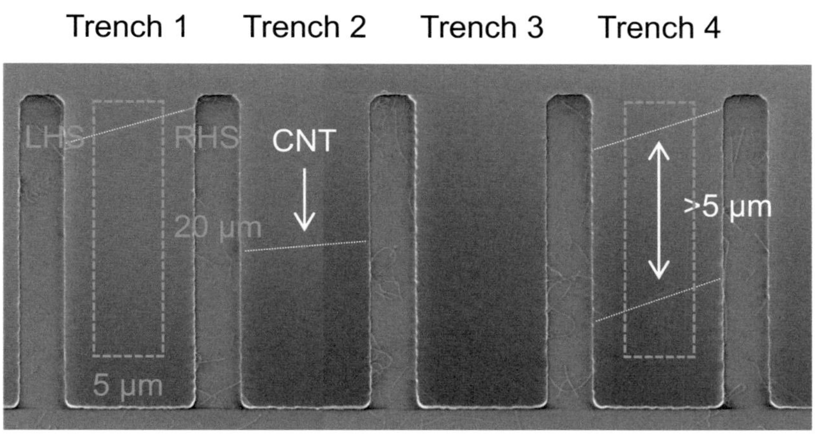

Figure 4.2: Graphical illustration of the definition of transferrable, individual CNTs. A hypothetical 5 μm-by-20 μm box is used for the definition.

Optimizing various growth parameters simultaneously with a full-factorial design of experiments (DoE) would be very time-consuming and require an excessive number of experimental runs. Instead, this work utilizes a multi-parameter optimization approach employing an orthogonal array, complemented by methodologies adopted from the Taguchi method – a strategy renowned in quality engineering for its efficiency in optimizing outcomes with a significantly reduced experimental runs [148]. This method has been used for the growth optimization of carbon nanotubes in other forms [149, 150]. It has been selected over the Box-Behnken design, another widely used DoE strategy, because the latter allows only three levels per parameter, which can constrain the resolution and accuracy of the results. One limitation to note about the orthogonal array-based Taguchi DoE is that the interplay between parameters is not taken into consideration.

A 4-level, 4-parameter orthogonal array DoE was designed, totaling 16 experiments instead of 256 experiments that would be required by a full-factorial DoE. The varied parameters included growth temperature, dilution ratio of ferritin solution, pressure, and CH_4 flow rate. The dilution ratio of ferritin solution is expected to directly affect the catalyst concentration on the substrate surface. These parameters and their levels were selected based on commonly used values in the literature [151–153] and existing know-how at the Micro and Nanosystems group at ETH Zurich. Operational constraints of the CVD oven (Aixtron BM 6 Pro), especially the temperature and gas flow rate limits, were also considered.

For example, originally, 100 sccm, 250 sccm, 400 sccm, and 550 sccm were desired levels for CH_4 flow rate. However, due to equipment limitations, the highest level was capped at 500 sccm. However, this adjustment does not impact the analysis of the results as the levels of each parameter do not require equal spacing.

The other experimental conditions besides the four selected parameters were kept the same. Catalyst deposition was performed as described in Section 3.2.2. Ferritin solutions were prepared by diluting a segment taken from an ultracentifuged stock solution [128] (Solution ID: 2015Sol2-3) by a required ratio with deionized water. 160 μl of diluted ferritin solution was used for each substrate. The growth duration was 30 minutes and the reduction step with H_2 was 2 minutes. Reduction was performed at the same temperature as CNT growth in each CVD run. Based on previous findings, the H_2 flow was maintained at a 2:3 ratio with CH_4, a proportion that had proven effective in earlier experiments conducted at the Micro and Nanosystems group. This means that a change in the CH_4 flow rate entails a proportional change in the H_2.

To evaluate the outcomes of these experiments, scanning electron microscopy (SEM) was used to analyze a total of 70 trenches from each CVD run, providing detailed insights into the CNT growth result using the definition of a transferrable, individual nanotube introduced earlier.

Table 4.2 summarizes the 4-level, 4-parameter orthogonal array DoE and its results. As shown in the table, each level of a parameter is used 4 times.

Table 4.2: 4-level, 4-parameter orthogonal array design of experiment (DoE) used in this work. Number of transferrable CNTs is based on SEM scans of 70 trenches per CVD run.

Run	Growth temp. (°C)	Ferritin solution dilution ratio	CH_4 flow rate (sccm)	Pressure (mbar)	No. transferrable CNTs
1	800	1:80	100	400	1
2	800	1:60	250	500	1
3	800	1:40	400	600	4
4	800	1:20	500	700	2
5	825	1:80	400	500	0
6	825	1:60	500	400	0
7	825	1:40	100	700	31
8	825	1:20	250	600	16
9	850	1:80	500	600	0
10	850	1:60	400	700	37
11	850	1:40	250	400	1
12	850	1:20	100	500	33
13	875	1:80	250	700	2
14	875	1:60	100	600	0
15	875	1:40	500	500	16
16	875	1:20	400	400	20

Figure 4.3 summarizes the mean number of transferrable CNTs at each level of a parameter. As shown in the table, each level of a parameter is used 4 times. Temperature and ferritin solution dilution ratio are found to have stronger influence on the growth result than pressure and CH_4 flow rate.

Analysis of each parameter provides further insights. The number of transferrable CNTs rises as temperature increases from 800 °C to 850 °C, after which further temperature increases lead to a decline in the number of transferrable CNTs. A closer look at SEM images from the experiments suggests that this decline is not only limited to transferrable CNTs but also overall growth activity. In other words, substrates from CVD runs at 875 °C have generally less CNTs both in the trenches and on the substrate surface. This trend can be explained by two opposing mechanisms. Higher reaction temperatures accelerate the endothermic dehydrogenation of methane, which is favorable for CNT formation. However, the exothermic crystallization of carbon into CNTs is inhibited by these higher temperatures. Consequently, at a certain temperature, the beneficial increase in reaction rate is outweighed by the slowdown of crystallization. Additionally, higher temperatures may increase the mobility of the catalyst, potentially leading to its loss from the substrate.

A lower dilution ratio of the ferritin solution leads to a higher number of CNTs, as anticipated. A higher concentration of ferritin results in a greater Fe catalyst concentration on the substrate, thereby enhancing CNT growth. However, excessive growth activity also increases the chance of growing bundled nanotubes or individual nanotubes that are too close to each other. A qualitative analysis of SEM images suggests that, beyond a certain concentration threshold, further increases in ferritin concentration might reduce the number of transferrable CNTs. There will be additional remarks on this point after a brief discussion about the two remaining parameters.

The influence of pressure on CNT growth does not present a clear trend, yet the highest mean number of transferrable CNTs was observed at the high pressure level of 700 mbar. Higher pressure increases the number of gas molecules in the chamber, thereby increasing the likelihood of carbon deposition. This increased probability of deposition at higher pressures is likely responsible for the enhanced CNT growth observed at the 700 mbar setting.

No definitive trend was observed for the effect of CH_4 flow rate on CNT growth. It is possible that two opposing effects contribute to this observation. A comparison between CVD Run 3 and Run 4 shows that increasing pressure, ferritin concentration, and flow rate led to a decrease in CNTs and overall growth activity. Given that higher ferritin concentration – and potentially higher pressure – generally enhances the count of transferrable CNTs, it can be inferred that an increase in CH_4 flow rate might negatively impact CNT growth. One possible explanation is that a higher gas flow rate reduces the residence time of gas molecules in the reaction chamber, thus decreasing the opportunity for CNT growth.

Although the results in Figure 4.3 suggest that the optimal set of parameters should be a growth temperature of 850 °C, a dilution ratio for the ferritin solution of 1:20, a pressure of 700 mbar, and a CH_4 flow rate of 400 sccm, a

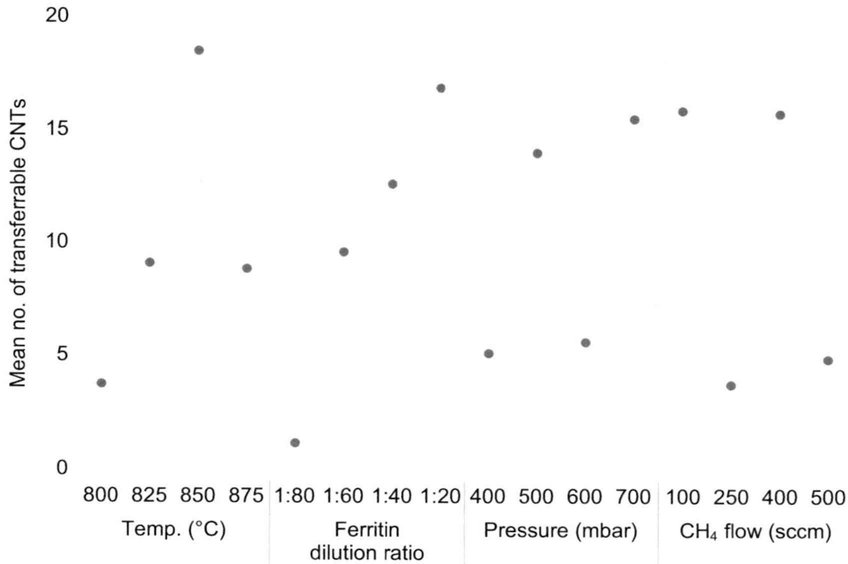

Figure 4.3: Mean number of transferrable CNTs for 4 parameters: temperature, ferritin solution dilution ratio, pressure, and CH_4 flow rate.

qualitative analysis of the SEM images from Run 10 and Run 12 indicates that further optimization beyond these runs was unnecessary or even not desirable.

This optimal set of parameters implies changing the ferritin solution dilution ratio from 1:60 to 1:20 in Run 10, while maintaining the other parameters the same. SEM images from Run 10 (Figure 4.4a) reveal that the substrate was generally very dense. Although many suspended nanotubes met the the aforementioned criteria for transferrable CNTs, only 3 out of 70 scanned trenches were single-CNT trenches (~4 %), and in multiple-CNT trenches, many suspended nanotubes were barely further than 5 μm from neighboring suspended CNTs. Consequently, it was concluded that altering the ferritin solution concentration to 1:20 was not a desirable approach.

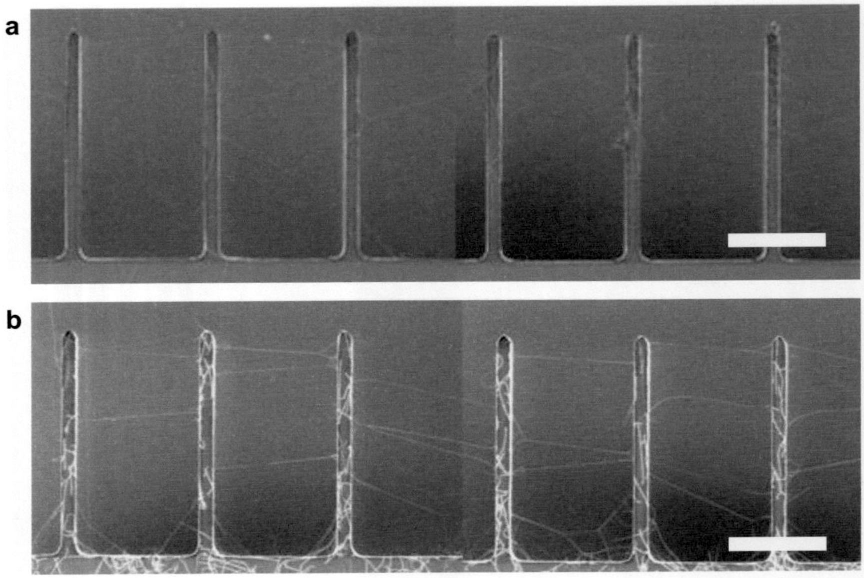

Figure 4.4: (a) SEM images from Run 12 (Sample ID: RB_Sample10_Fork2). (b) SEM images from Run 12 (Sample ID: RB_Sample10_Fork1). Scale bars: 10 μm. SEM images courtesy of Roland Bockholt.

Run 12 also provided interesting insights. SEM images from Run 12 (Figure 4.4b) displayed a very high number of nanotubes overall. The growth temperature of 850 °C, pressure of 500 mbar, and CH_4 flow rate of 100 sccm appeared to be highly effective – more so than the parameters used in Run 10 – in activating Fe catalyst particles, as evidenced by the number of nanotubes on the substrate surface. However, only 4 out of 70 trenches were single-CNT trenches (~6 %). Therefore, an additional experiment was conducted with the same temperature, pressure, and CH_4 flow rate as Run 12, but with a 1:40 dilution ratio of the ferritin solution.

As shown in Figure 4.5, this adjusted set of parameters reduced the overall growth activity as expected and increased the ratio of single-CNT trenches to 29 % (20 out of 70 trenches). Thus, the growth parameters of 850 °C temperature, 500 mbar pressure, 100 sccm CH_4 flow rate, and 1:40 dilution ratio for ferritin solution were determined to be the set of parameters for the remainder of this thesis.

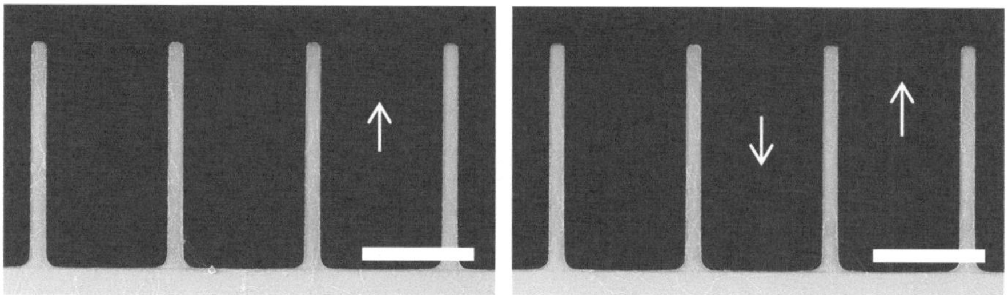

Figure 4.5: SEM images from a CVD growth run with the final set of parameters – 850 °C temperature, 500 mbar pressure, 100 sccm CH_4 flow rate, and 1:40 dilution ratio for ferritin solution (Sample ID: SJGS212-Fork1). For better visibility, suspended CNTs are marked with arrows. Scale bars: 10 μm.

4.2.3 Large-scale growth substrate

In parallel to the optimization of suspended CNT growth parameters, new large-scale growth substrates were designed and fabricated [126]. This approach aims to significantly increase the hosting capacity of each growth substrate. For clarity, the growth substrate described so far (Figure 3.3 will be referred to as a *small-scale growth substrate* from this point onward, and the new one, a *large-scale growth substrate*. The large-scale growth substrate shares the same fabrication process as the small-scale substrate, as described in Section 3.2.1.

The small-scale growth substrates are square in shape and contain 16 forks, each with 35 trenches, totaling 560 trenches per substrate. The overall dimension is approximately 3 mm by 3 mm. The large-scale substrate, depicted in Figure 4.6a-b, is a disk with a diameter of 39 mm and it features 64 forks and a total of 5888 trenches (92 trenches per fork). The dimensions of this substrate were chosen based on several factors primarily related to the robotic assembly system used for the mechanical transfer of CNTs onto devices.

The catalyst deposition process starts with the deposition of a diluted ferritin solution onto the substrate. For the small-scale substrates, this is done by fully submerging them in the ferritin solution for 4 minutes. However, this approach is inefficient for the large-scale substrates, as it would result in excessive waste of ferritin solution on the substantial substrate body part without any forks. To address this issue, a new ferritin solution deposition tool has been developed. On this tool, the disk-shaped substrate is fixed on a rod and only a few forks of the large-scale substrate is immersed in 350 μl of ferritin solution (Figure 4.6c). Then, the substrate is rotated at

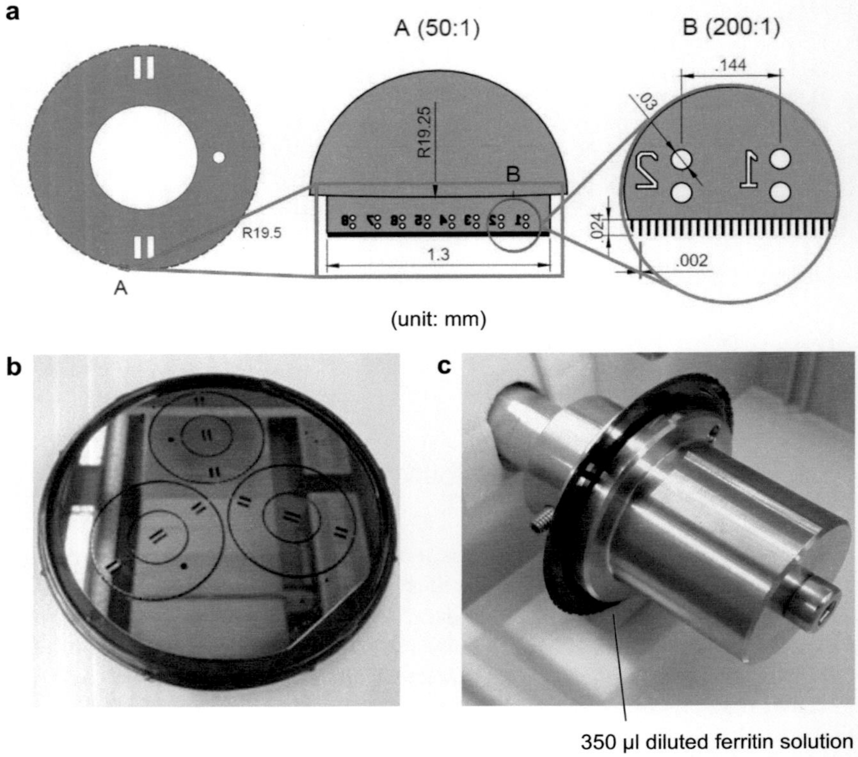

Figure 4.6: a) Graphical illustration of a large-scale growth substrate [126]. The disk-shaped substrate consists of 64 forks on its circumference and approximately 6000 trenches. (b) Picture of 3 large-scale growth substrates fabricated on a 4-inch silicon-on-insulator (SOI) wafer. (c) A large-scale growth substrate mounted on the ferritin solution deposition tool [142].

the speed of 7.5 revolutions/minute for 10 minutes. Considering that the small-scale substrate, which hosts 10 times less trenches, requires about half of the amount (160 μl), this new tool contributes significantly to efficient use of ferritin solution. This tool was designed and built by Julian Radziwill, whose research project was supervised by the author [142].

4.2.4 Batch growth on large-scale substrates

This section demonstrates the batch growth of suspended carbon nanotubes (CNTs) by combining the results from the two previous sections. In this phase of the study, the optimal set of growth parameters identified in Section 4.2.2 – 850,°C temperature, 500 mbar pressure, 100 sccm CH_4 flow rate, and a 1:40 dilution ratio for the ferritin solution – was employed to grow CNTs on the large-scale growth substrates described in Section 4.2.3.

Part of the experimental work reported in this section was conducted by Kristian Ivkovic, whose bachelor thesis was supervised by the author. The author prepared the large-scale growth substrates and ferritin catalyst solutions used in the experiments and provided guidance on the design of experiments. Kristian Ivkovic performed the CVD growth runs and conducted the SEM scans. This section summarizes the key concepts and results of the study, but it does not delve into all the details of statistical analysis. For details beyond what is available in this work, the reader is referred to Kristian Ivkovic's report [143], which can be provided by the author upon request.

Three CVD runs were conducted, with each run containing three large-scale substrates placed on a silicon support wafer in the CVD oven as depicted in Figure 4.7b. The forks on the device layer of the silicon-on-insulator (SOI) wafer do not directly touch the support wafer. Depending on their position in the CVD oven, they are named Top, Left, and Right substrates. According to the manufacturer, Aixtron, the CVD oven is designed to maintain uniform parameters across the central 4-inch area of the system. Therefore, the placement of three large-scale substrates was intended to distribute the forks as evenly as possible across this 4-inch area and to assess the uniformity of the growth results across this region.

After the CVD runs, four forks were scanned by SEM on each substrate. These four forks were located at 0°, 90°, 180°, and 270° positions relative to the indexing hole on the substrate as illustrated in Figure 4.7a. Scanning these specific forks after the CVD runs enabled tracking of the growth results across the central 4-inch area of the system. It also allowed comparing the growth results from run to run at approximately the same positions within the CVD oven, thereby providing a consistent basis for evaluating the uniformity and reproducibility of the CNT growth across the CVD runs.

Figure 4.8, which mimics the actual positions and orientation of the three substrates in the CVD oven, shows the distributions of no-CNT, single-CNT, and multiple-CNT trenches across the 12 SEM-scanned forks. A first glance at the figure immediately suggests a few noticeable trends.

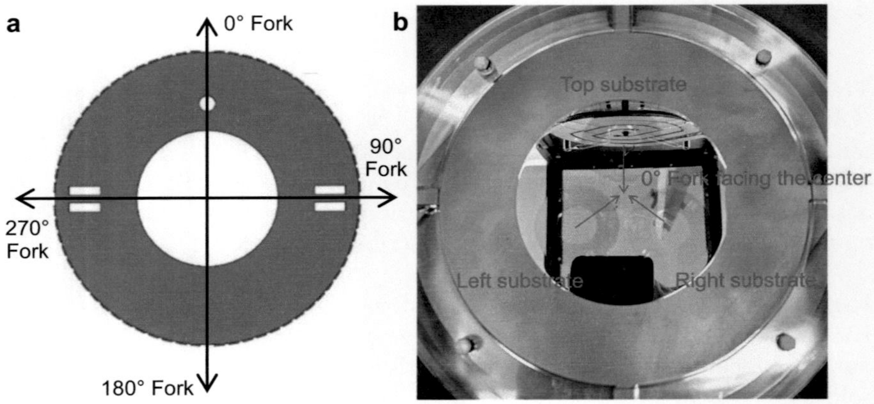

Figure 4.7: a) Graphical definition of 0°, 90°, 180°, and 270° forks on each large-scale substrate. This is a re-created drawing of an original illustration from Kristian Ivkovic. (b) In the CVD growth runs, three large-scale growth substrates – each named Top, Left, and Right substrate – were placed in the CVD oven as shown in the picture. Picture courtesy of Kristian Ivkovic.

First, CNT growth was significantly more active on the left side of the CVD oven, as indicated by the very high number of multiple-CNT trenches on the left half. In case of the 90° Fork of the Left substrate, 76 trenches out of 92 (83 %) had multiple CNTs. Second, CNT growth was not very active in the center of the CVD oven, represented by the high count of no-CNT trenches on the 0° forks of all three substrates. These trends are quantitatively confirmed with the counts of single-CNT trenches and multiple-CNT trenches shown in Table 4.3 and Table 4.4.

Table 4.3: Counts of single-CNT trenches in the first CVD run.

Substrate	**Top**	**Left**	**Right**
0° Fork	12	6	7
90° Fork	27	9	26
180° Fork	24	14	28
270° Fork	18	19	24
Average	**17.8**	**Standard Deviation**	**7.7**

The Left substrate showed an average of 47.8 multiple-CNT trenches per

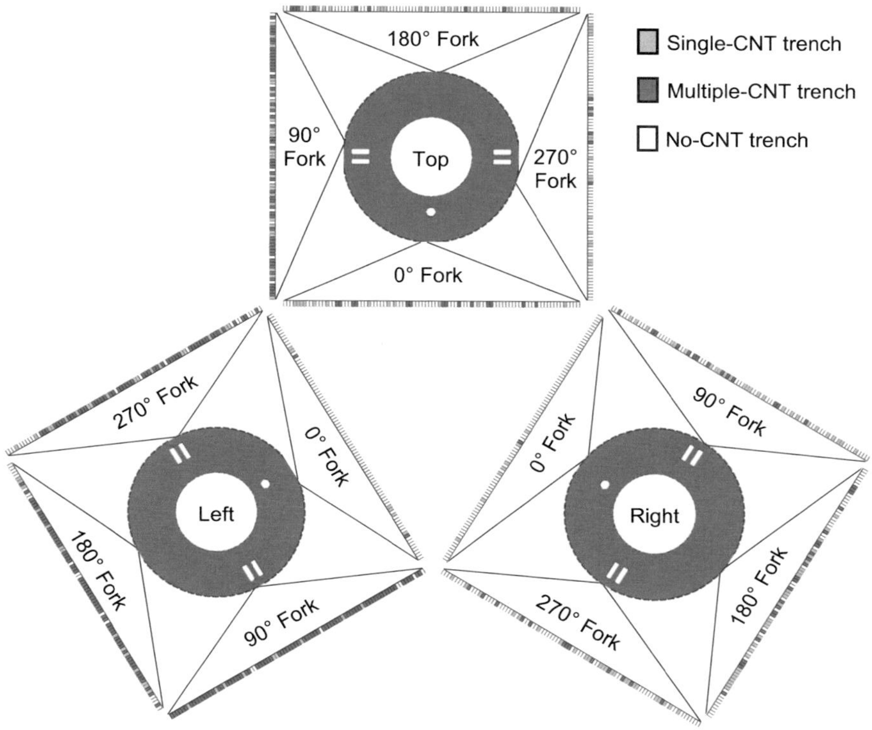

Figure 4.8: Distribution of no-CNT trenches, single-CNT trenches, and multiple-CNT trenches in the first CVD run. This is a re-created drawing of an original illustration from Kristian Ivkovic.

fork, which is 35 trenches more than the right substrate, which had an average of 12.8 multiple-CNT trenches per fork. In all three substrates, the 0° fork had the least number of single-CNT and multiple-CNT trenches. This disparity highlights the significant non-uniformity in CNT growth depending on the position within the CVD oven. Some of the possible explanations for the non-uniformity are as follows:

- The temperature control is performed with an infrared sensor that detects the temperature of only the center point of the support wafer. Below the 4-inch support wafer are a graphite cathode and a heater element, both with a diameter of 6 inches. Temperature control based on only one spot might not ensure uniform temperature distribution across the 4- or 6-inch areas.

Table 4.4: Counts of multiple-CNT trenches in the first CVD run.

Substrate	Top	Left	Right
0° Fork	14	2	1
90° Fork	51	76	14
180° Fork	9	61	23
270° Fork	6	52	13
Average	**26.8**	**Standard Deviation**	**26.4**

- Gases are injected into the chamber through a showerhead gas delivery system with multiple holes, but the gas flow rates in different directions are not measured separately. Therefore, different areas in the chamber might experience different gas flow rates.

Focusing solely on the count of single-CNT trenches might lead to an oversight of the position-dependent non-uniformity of CNT growth. Table 4.3 indicates a standard deviation of 7.7 in the number of single-CNT trenches across all 12 SEM-scanned forks, which, on the surface, may not seem very high. Therefore, counting both single-CNT trenches and multiple-CNT trenches was crucial in clearly identifying the position-dependent non-uniformity of CNT growth activity.

Figure 4.9, 4.10, and 4.11 show SEM images from the scanned forks presented in Figure 4.8. Note that, as described already, the forks on the same growth substrate show drastically different growth activities. The number of suspended CNTs, which are marked with arrows, is a good indicator of CNT growth activity, but one can see more drastic differences in the number of substrate-bound CNTs on the fork body area, an example of which is marked with a box in Figure 4.9. For example, a very small number of CNTs – both substrate-bound and suspended – are to be found on the 0° Forks of all three substrates, whereas one can see very densely covered fork body areas from the 90° Fork and 180° Fork of the Left substrate, among others.

These trends were consistently observed in the second and third CVD runs, the data for which is presented in the Appendix C. Across the 9 substrates and 36 scanned forks, a total of 758 single-CNT trenches and 803 multiple-CNT trenches were identified. On average, this equates to 22.0 single-CNT trenches per fork (24 %), which is comparable to the ratio of

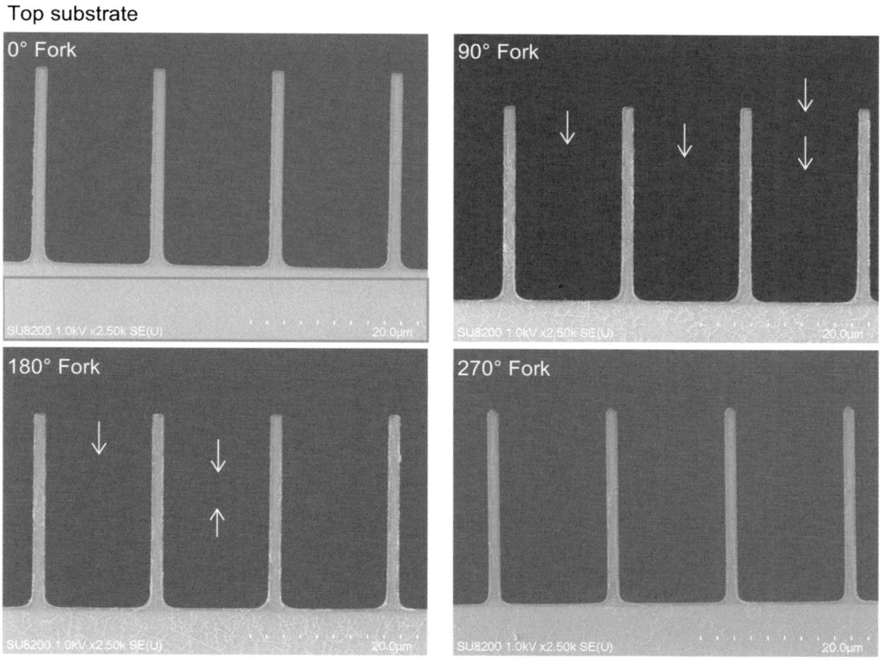

Figure 4.9: SEM images of the four scanned forks of the Top substrate in Figure 4.8. Suspended CNTs are marked with arrows, and the fork body area on the 0° Fork is marked with a box. SEM images from Kristian Ivkovic.

single-CNT trenches observed on the small-scale growth substrates after the multi-parameter optimization of suspended CNT growth (Section 4.2.2).

A simple extrapolation suggests that up to 1400 single-CNT trenches per substrate and 4200 single-CNT trenches per CVD run could be achievable. However, the observed position-dependent non-uniformity underscores the necessity for collecting more data points from a larger number of forks in the CVD runs. While the data from this study suggests that the non-uniformity is primarily attributable to the position of the fork within the CVD oven, other factors such as non-uniform coverage of catalyst particles cannot be entirely ruled out.

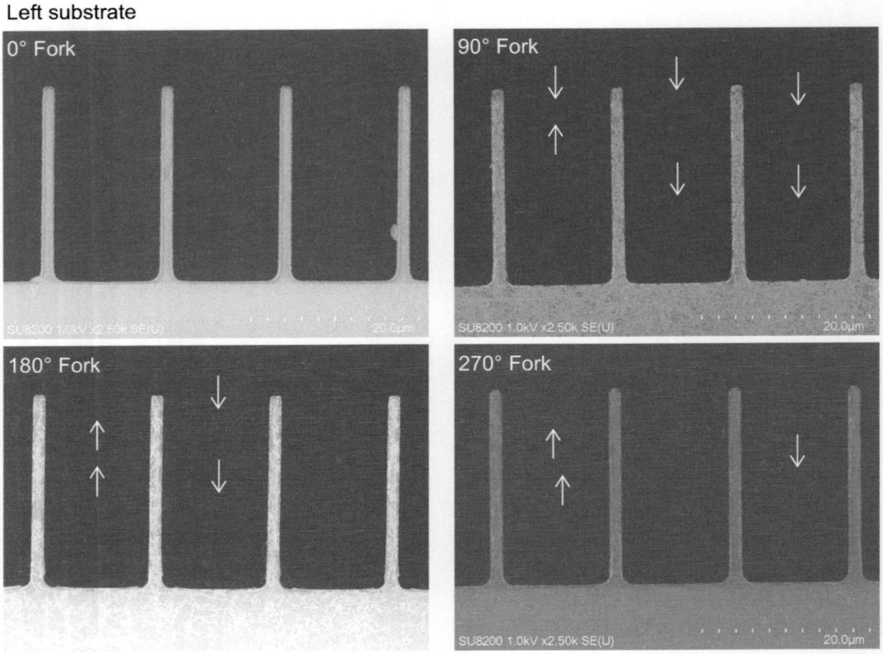

Figure 4.10: SEM images of the four scanned forks of the Left substrate in Figure 4.8. Suspended CNTs are marked with arrows. SEM images from Kristian Ivkovic.

4.3 Automated robotic assembly of carbon nanotubes

4.3.1 Automated nanoassembly machine

Earlier in Section 2.2.2, the concept of a fully automated, robotic assembly system for individual carbon nanotubes (CNTs) was introduced. This section discusses the realization of such a system and the devices assembled by it. The system, referred to as the automated nanoassembly machine from this point onward (Figure 4.12), is a parallel kinematic micromanipulator that employs precision engineering and machine vision to perform multiple transfers of carbon nanotubes from growth substrates onto device substrates in an iterative manner, as depicted in Figure 2.11. The nanoassembly machine is designed to be compatible with wafer-level assembly on 4-inch wafers, which can be covered by the x-y motion range

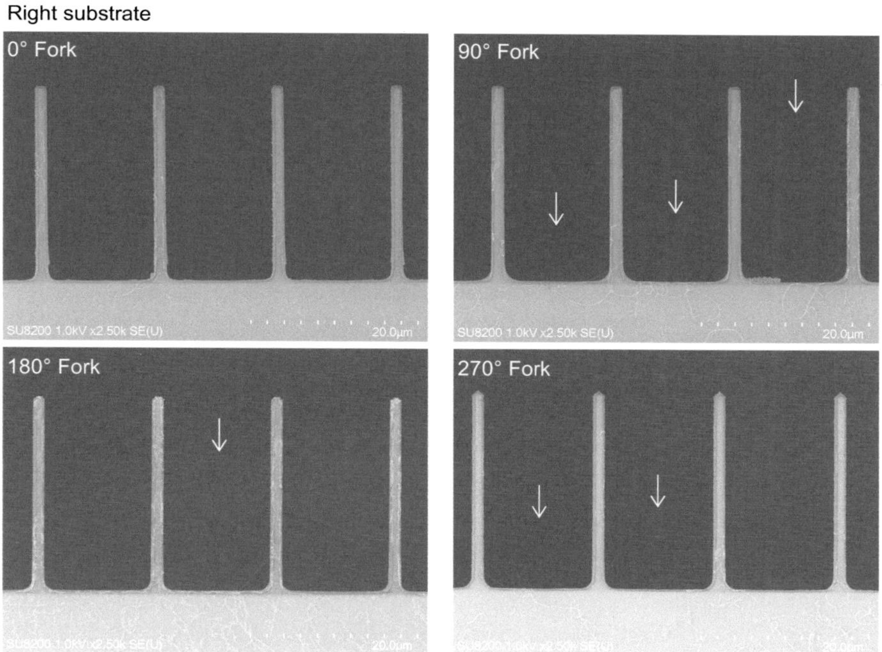

Figure 4.11: SEM images of the four scanned forks of the Right substrate in Figure 4.8. Suspended CNTs are marked with arrows. SEM images from Kristian Ivkovic.

of the machine. The system maintains its motion accuracy over a large area by recalibrating the positions of growth substrate and device substrate when its traveling distance exceeds a certain value.

Since the detailed development and implementation of this system are beyond the scope of this thesis, readers are kindly referred to the comprehensive work by Butzerin *et al.* [154] for an in-depth exploration of the automated nanoassembly machine.

4.3.2 Electrical characterization of assembled devices

In this section, the focus moves to the electrical characterization results of the suspended CNTFETs assembled by the automated nanoassembly machine. The electrical properties of these devices provide crucial insights into the performance and reliability of the fully automated assembly process.

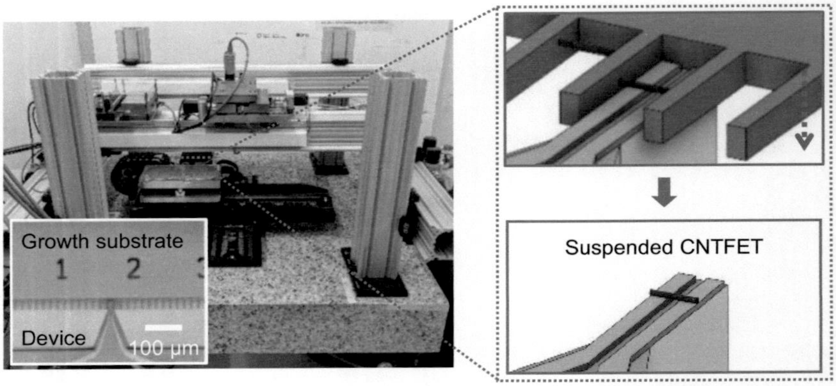

Figure 4.12: Automated nanoassembly machine. Inset: microscopic image of the nanoassembly process that transfers a suspended carbon nanotube onto a device.

The general workflow for the experiments follows the descriptions provided in Section 4.1. Several hundred suspended carbon nanotubes were grown on large-scale growth substrates using the optimized growth parameters. These nanotubes were then scanned using Raman spectroscopy. Single-CNT trenches were selected according to the definition introduced earlier. This list of single-trench trenches was subsequently fed to the automated nanoassembly machine, which transferred the nanotubes from the selected single-CNT trenches onto the device substrates. To ensure clean electrical contact between the nanotubes and device electrodes, the contact improvement methods described in Chapter 3 were performed before and after the mechanical CNT transfers. After thermal annealing, which is the post-transfer contact improvement method, the devices were measured electrically. The experimental support of Natanael Lanz, Andre Butzerin, and Marko Nikolic in the operation of the automated nanoassembly machine and Maria El Abbassi, Luca Ornago, and Frederik van Veen in device fabrication, Raman spectroscopy, and electrical measurements is kindly appreciated.

It is important to note that the device substrates used in these experiments had smaller source-drain distances (L_{sd}) of 0.5 μm (500 nm) and gate lengths (L_{gate}) of 0.12 μm (120 nm) than the devices described so far. These reduced dimensions necessitated the use of e-beam lithography for patterning the electrodes, instead of optical lithography. The device geometry is illustrated in Figure 4.13.

The devices were fabricated on a 4-inch wafer that was partially diced (top

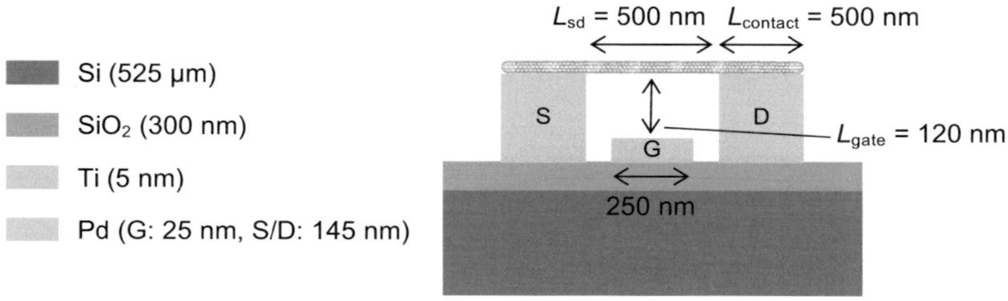

Figure 4.13: Graphical illustration of the 4-inch device wafer containing partially diced 1 cm-by-1 cm device chips. The chip blocks used in the three automated transfer sessions are marked.

300 μm) into 1 cm-by-1 cm chips. Each chip contains 10 devices, with 5 devices facing one direction and the other 5 facing the opposite direction. In the robotic assembly sessions, only the 5 devices facing the top direction on each chip were used. The partial dicing facilitated the cleaving of the chip blocks. The automated transfers were performed in three sessions, and the chip block used for each session is illustrated in Figure 4.14. For instance, in Session 1, a 3 cm-by-4 cm chip block located at the bottom-right side of the wafer was used, with its 40 devices facing the same direction serving as the receiving devices.

In total, 107 suspended CNTs were transferred onto devices in 140 minutes across three assembly sessions, achieving a transfer rate of 46 transfers per hour. Each session was run with little to no human intervention, which was limited to accepting the system software to proceed when the confidence level about the alignment between the growth substrate and device substrate, which was determined by the machine software, was lower than a defined level. Session 3 consisted of two device chip blocks, and the exchange of the chip block was done manually.

Out of the 107 transferred CNTs, 74 devices were electrically active, corresponding to an overall yield of 70 %. An electrically active device is defined by the presence of a measurable source-drain current (I_{sd}) without a short to the gate. The detailed results of the three assembly sessions are illustrated in Figure 4.15a.

A session-by-session breakdown is as follows. In Session 1, 40 transfers were completed in 55 minutes (44 transfers/hour). 30 devices were electrically active (75 %). In Session 2, 20 transfers were completed in 31 minutes (39 transfers/hour). 16 devices were electrically active (80 %). In Session 3, 47

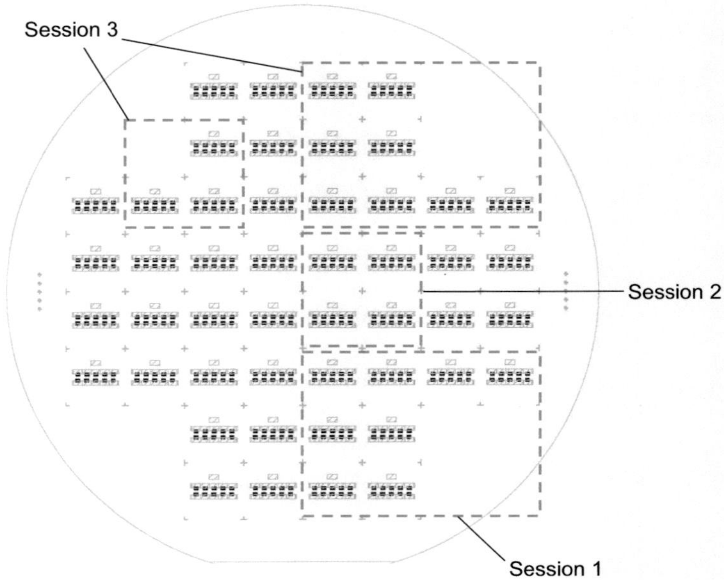

Figure 4.14: Graphical illustration of the 4-inch device wafer containing partially diced 1 cm-by-1 cm device chips. The chip blocks used in the three automated transfer sessions are marked.

transfers were completed in 53 minutes (53 transfers/hour). 28 devices were electrically active (60 %). These results demonstrate a consistent and high transfer rate, with a substantial percentage of active devices across different sessions. Transfer rate was also maintained constant across sessions.

The transfer characteristics (I_d-V_{gs}) of the assembled CNTFETs were analyzed to evaluate their performance. Figure 4.16 shows the transfer characteristics of all assembled CNTFETs, with two groups – semiconducting devices (Figure 4.16a) and metallic and small-gap semiconducting devices (Figure 4.16b) – plotted separately for better visibility. V_{sd} was set to 10 mV. This is lower than the value used for previously measured devices with L_{gate} of 1 μm (500 mV). This was intentional because L_{gate} of these devices was significantly smaller (120 nm) and lower biases were needed to avoid "collapses" of the suspended CNTs onto gates. For the same reason, the V_{gs} sweeps were kept to smaller ranges, mostly [-2 V, 2 V]. For the categorization of semiconducting, metallic, and small-gap semiconducting CNTs, the criteria described in Section 3.3.1 were followed.

Overall, 33 devices were identified as semiconducting (45 %), 5 devices metallic (7 %), 36 devices small-gap semiconducting (49 %). Figure 4.15b presents the distribution of semiconducting, metallic, and small-gap semiconducting CNTs across the three sessions. The unusually high ratio of small-gap semiconducting CNTs observed could be attributed to several factors. First, it has been shown that thermal annealing consistently opens a gap in mechanically transferred, suspended CNTs [115]. Second, Section 4.2.4 suggests that the CVD growth of suspended nanotubes is highly non-uniform across the forks on the same growth substrate. Although the uniformity analysis of Section 4.2.4 only focuses on the number of suspended CNTs and overall growth activity, not the types of CNTs, it is possible that the forks used for these assembly sessions had an unusually high ratio of small-gap semiconducting CNTs. This growth non-uniformity might explain why more than half of the small-gap semiconducting devices were fabricated in Session 1. Lastly, the downward transfer motion of the nanoassembly machine was set at a slow speed of 1 μm/second. This slow transfer motion might have induced strain in the nanotube.

Table 4.5: Categorization of the automatically assembled CNTFETs based on electrical characteristics.

	Semiconducting	**Metallic**	**Small-gap semiconducting**
Session 1	9 (30 %)	1 (3 %)	20 (66 %)
Session 2	9 (56 %)	2 (13 %)	5 (31 %)
Session 3	15 (54 %)	2 (7 %)	11 (39 %)
Total	**33 (45 %)**	**5 (7 %)**	**36 (49 %)**

The automatically assembled can be compared to manually assembled devices in terms of key performance metrics. The manual assembly was described in Section 3.2.5 and the resulting devices were analyzed in Chapter 3.

Figure 4.17 presents the empirical cumulative distributions of R_{on} for all devices (Figure 4.17a) and semiconducting devices only (Figure 4.17b). The R_{on} values are defined at V_{gs} = -2 V or at the lowest V_{gs} measured. Both groups of devices underwent the same contact improvement protocols, as described in Sections 3.2.4 and 3.2.6. The manually assembled devices shown in Figure 4.17 correspond to the Group 3 devices in Figure 3.9. In Figure

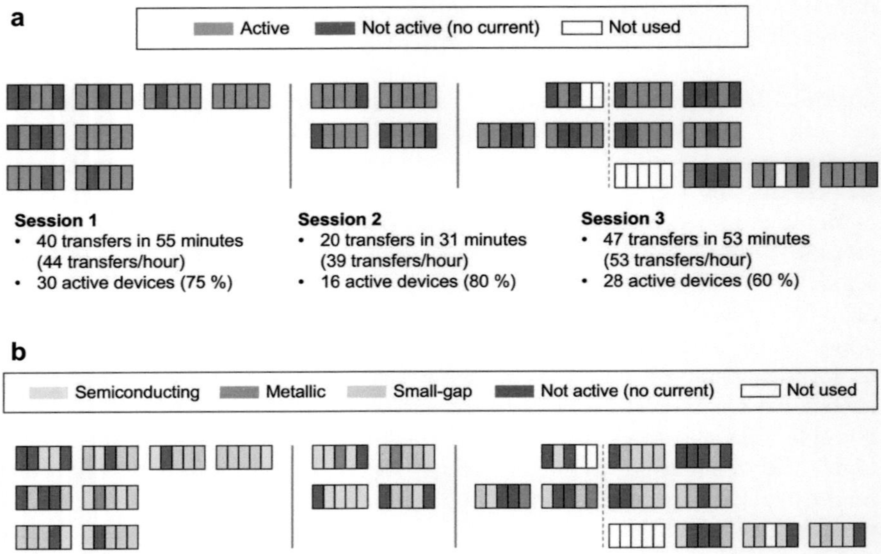

Figure 4.15: (a) Distribution of electrically active, not active, and not used devices across three sessions of automated CNT transfer. (b) Classification of device types across the sessions: semiconducting, metallic, small-gap semiconducting, not active, and not used. CNT transfer was not attempted on the *not used* devices because their electrodes had been found to be damaged or not clean by optical microscopy.

3.9, R_{on} was defined at V_{gs} = -10 V. However, in Figure 4.17a, their R_{on} values are defined at V_{gs} = -2 V for a more straightforward comparison with the automatically assembled devices.

Figure 4.17a shows that the automatically assembled devices have a median value of 135 kΩ with an interquartile range (IQR) of 1.47 MΩ. This median value is lower than that of the manually assembled devices, which have a median R_{on} of 306 kΩ. However, the IQR for the automated assembly is significantly larger than that of the manually assembled devices, which have an IQR of 235 kΩ.

Among the automatically assembled devices, many are metallic and small-gap semiconducting devices. On the other hand, the group of manually assembled devices contains only two metallic and small-gap semiconducting devices. Therefore, the semiconducting devices from both groups are compared separately as shown in Figure 4.17b. Analyzing the

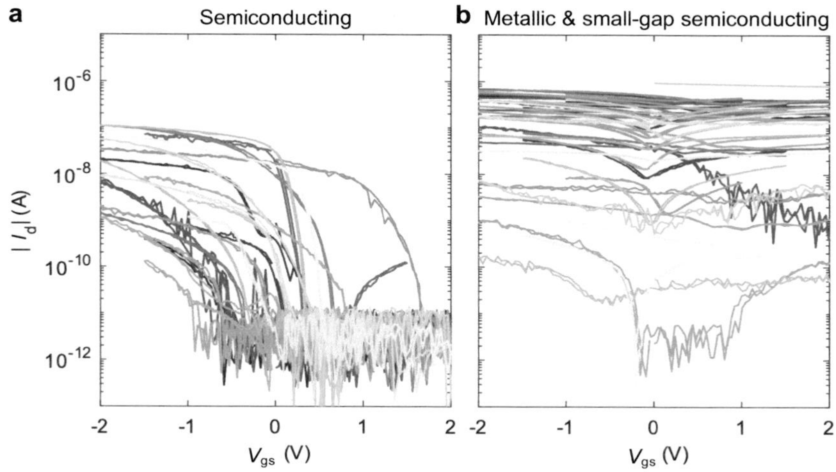

Figure 4.16: Transfer characteristics (I_d-V_{gs}) of CNTFETs assembled by the automated nanoassembly machine measured at V_{sd} = 10 mV. (a) Semiconducting devices. (b) Metallic and small-gap semiconducting devices.

semiconducting devices separately also allows the investigation of V_{th} and SS, as was done in Figure 3.10.

Figure 4.18a and b compare V_{th} and SS of manually assembled and automatically assembled semiconducting devices, respectively. In case of V_{th}, the manually assembled devices exhibit a median of 1.62 ± 0.72 V (IQR), whereas the automatically assembled devices show a median V_{th} of 0.05 ± 0.79 V.

Regarding SS, the manually assembled devices have a median SS of 200 ± 104 mV/dec, while the automatically assembled devices have a median SS of 180 ± 145 mV/dec.

For both V_{th} and SS, two-sample K-S tests indicate that the null hypothesis is rejected, with low p-values of 2.71e-5 for V_{th} and 0.04 for SS.

However, before concluding that these differences in V_{th} and SS are solely due to the different CNT assembly methods, it is important to consider that V_{th} and SS are also significantly influenced by device architecture and geometry. Figure 3.10 already showed that distinct groups of devices with

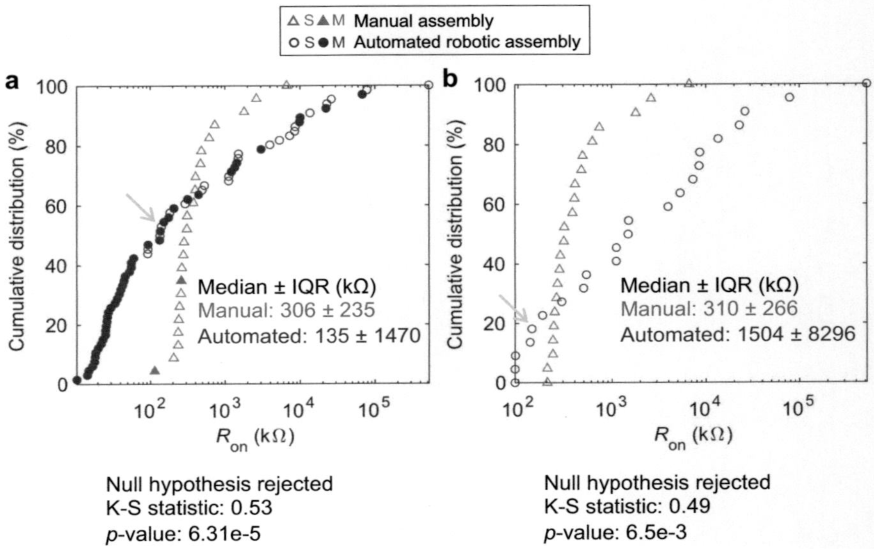

Figure 4.17: Empirical cumulative distributions of R_{on} for (a) all devices and (b) semiconducting devices assembled manually (triangles) and automatically (circles). R_{on} is defined at V_{gs} = -2 V. If V_{gs} sweep of a device does not include -2 V, R_{on} is defined at the lowest V_{gs} measured. In (a), empty markers represent semiconducting devices, and filled markers represent metallic and small-gap semiconducting devices. The device, which is tested as a gas sensor in Section 5.3.4, is marked with an arrow. The results of two-sample Kolmogorov-Smirnov tests are summarized on the bottom.

very different R_{on} can exhibit similar distributions of V_{th} and SS, provided that the device groups share the same architecture and geometry.

The manually assembled devices and automatically assembled devices compared here have different geometrical dimensions. As shown in Figure 3.2f, the manually assembled devices have $L_{\text{gate}} = 1\ \mu$m, $L_{\text{sd}} = 2.8\ \mu$m, and $L_{\text{contact}} = 2\ \mu$m. In contrast, the automatically assembled devices have $L_{\text{gate}} = 120$ nm, $L_{\text{sd}} = 500$ nm, and $L_{\text{contact}} = 500$ nm, as shown in Figure 4.13.

These differences in geometrical dimensions are expected to affect SS, which is defined as:

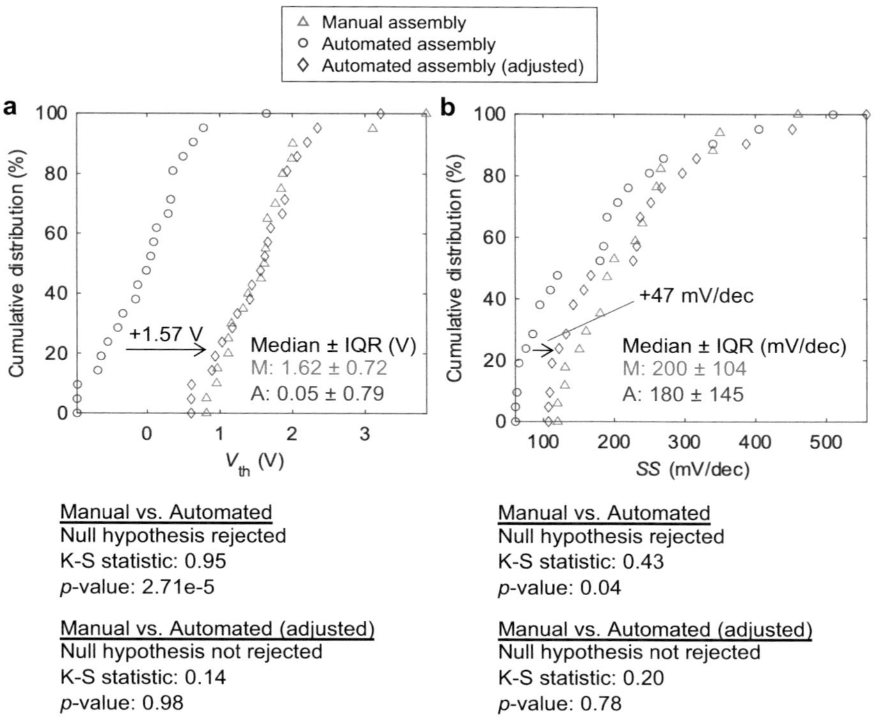

Figure 4.18: Empirical cumulative distributions of (a) threshold voltage (V_{th}) and (b) subthreshold swing (SS) of semiconducting devices assembled manually (circles) and automatically (triangles). The third group of devices, which results from an adjustment of geometrical or environmental factors, is represented with gray diamonds. The results of two-sample Kolmogorov-Smirnov tests are summarized on the bottom.

$$SS = \ln(10)\frac{kT}{q}\left(1 + \frac{C_d}{C_{\text{ox}}}\right) \tag{4.1}$$

where C_d is the depletion layer capacitance, C_{ox} is the gate dielectric capacitance, and kT/q is the thermal voltage. The subthreshold slope (SS) depends on these capacitances, which are influenced by the device geometry. While C_d can be affected by individual CNTs, C_{ox} is largely determined by the device geometry, especially L_{gate}. Therefore, it is crucial to account for these geometrical differences when comparing SS of manually and automatically assembled devices. Equation 4.1 predicts an offset of 47 mv/dec from the geometrical differences between the two groups of devices. Therefore, a third group of data points has been created by adding 47 mV/dec to the SS values of automatically assembled devices. This third set is denoted as Automated (adjusted) in Figure 4.18b. With this adjustment, the K-S test suggests that the SS distributions of the Automated (adjusted) group and the Manual group are likely to belong to the same statistical distribution, with a p-value of 0.78.

A similar, additive adjustment has been applied to V_{th} to account for the difference in the measurement environment. The manually assembled devices were measured in cleanroom air with 40 % R.H., whereas the automatically assembled devices were measured in a vacuum prober. In this case, the difference between the median V_{th} of the manually assembled devices and the median V_{th} of the automatically assembled devices (1.57 V) has been added to the V_{th} values of the automatically assembled devices. As shown in Figure 4.18a, the K-S test suggests that the V_{th} distributions of the Automated (adjusted) group and the Manual group are, again, very likely to belong to the same statistical distribution, with a p-value of 0.98.

The K-S tests after the compensation of the geometrical and environmental differences seem to imply that the choice between manual CNT assembly and automated CNT assembly does not introduce a statistically meaningful difference to the distributions of V_{th} and SS. In Chapter 3, discussions about Figure 3.10 and Figure 3.9 suggested that the distribution of R_{on} is largely determined by contact quality.

With this in mind, the comparison of R_{on} distributions of manually assembled devices and automatically assembled devices can be revisited (Figure 4.17).

Among the automatically assembled devices, more than 60 % of devices show R_{on} below 240 kΩ even at V_{gs} = -2 V. The R_{on} of 240 kΩ is highlighted here because, for devices used in Chapter 3, the R_{on} value of 240 kΩ (at V_{gs} = -10 V) appeared to be the threshold between the devices that show an NDC behavior and those that break before reaching the NDC regime. The high number of low-resistance devices in the automated assembly group can be explained by the shorter L_{sd}. In case of CNTs grown by chemical vapor deposition (CVD), each μm of CNT accounts for a few tens of kΩ [113,114]. The L_{sd} difference between the two groups is 2.3 μm, which corresponds to a resistance difference of 50-100 kΩ.

However, the high number of devices with $R_{on} > 1$ MΩ (33 %) is out of expectation because all of the devices were assembled within 1 hour from electrode cleaning with Ar-ion. This implies a chance that the contact quality of the devices in this group was not very consistent. In fact, when these devices were fabricated, the e-beam evaporation step for the Pd electrodes saw unusually severe fluctuations in Pd deposition rate. Several batches of devices, which were fabricated around the same time as the devices used in this study, could not be used because of this issue, and in severe cases, Pd electrodes showed abnormal spikes and were easily delaminated. It is possible that the usual Ar-ion cleaning and thermal annealing parameters were not sufficient for these suboptimal Pd electrodes. However, at this point, the chance that the automated CNT assembly increases the distribution of R_{on} cannot be ruled out either. To conclude whether the high-R_{on} devices are the outcome of suboptimal device substrate fabrication or CNT assembly method, further investigations about the contact conditions of these device substrates with XPS and AFM could be helpful. Manual transfer of CNTs onto these device substrates could also provide a direct point of comparison.

4.4 Chapter summary

This chapter presented discussions about the batch growth of suspended carbon nanotubes (CNTs) and their subsequent automated assembly with a robotic machine.

Initially, the optimization of CNT growth parameters was addressed. The goal was to maximize the number of transferrable, individual CNTs. Through a systematic multi-parameter optimization process, the optimal growth conditions were identified: a growth temperature of 850 °C, a pressure of 500 mbar, a CH_4 flow rate of 100 sccm, and a ferritin solution dilution ratio of 1:40. This optimized set of parameters significantly improved the yield of transferrable CNTs.

Following the growth optimization, the development and implementation of large-scale growth substrates were discussed. These substrates were designed to host thousands of trenches, increasing the number of potential CNT growth sites. A novel catalyst deposition method was devised to ensure uniform and material-efficient deposition of ferritin on the forks of large-scale growth substrates.

Combining the results of these two strategies, batch growth of suspended CNTs was attempted on 3 large-scale growth substrates put in the same CVD run. Overall, 24 % of trenches were identified as single-CNT trenches with transferrable CNTs, equivalent to approximately 1400 single-CNT trenches per substrate and 4200 single-CNT trenches per CVD run. Nevertheless, significant non-uniformity in CNT growth activity was observed at different positions within the CVD oven.

The chapter then discussed the automated assembly of CNTs. The automated nanoassembly machine, a parallel kinematic micromanipulator, transfer pre-selected CNTs from growth substrates to device substrates. This system achieved a transfer rate of approximately 46 transfers per hour, with a 70 % yield of electrically active devices. The performance of this system is compared with other nanostructure integration methods in Table 2.4.

The electrical characterization of the assembled CNTFETs provided insights into the distribution of different types of devices, and also the distributions of R_{on}, V_{th}, and SS. The automatically assembled devices showed a comparable yet lower median R_{on} value than manually assembled devices, but the IQR was significantly higher.

The results demonstrated the efficacy of the automated robotic assembly of carbon nanotubes for the fabrication of ultraclean CNTFETs at a high throughput, highlighting its potential for scalable fabrication of CNT-based electronic devices.

5 Fast Sensor Operation at Ultra-low Power

This chapter reports fast and ultra-low-power operation of suspended carbon nanotube-NO_2 sensors. The proposed operation method is enabled by nanotube self-heating and transient sensing. As discussed in Section 2.1.4, the self-heating effect in the nanotube channel significantly accelerates the desorption of gas molecules, reducing the sensor recovery time to a few minutes. As gas molecules re-adsorb on the nanotube after cooling, the initial rate of the sensor transient is used to determine NO_2 concentration within a few minutes. This work considers temperature profiles along the self-heated carbon nanotube, their effect on different sensing regions, and a physical model-based fit for transient sensing. The content of this chapter is largely reported in [75].

A color version of this chapter is available at https://doi.org/10.3929/ethz-b-000702474.

5.1 Sensor operation principles

As described in Section 1.2.1, carbon nanotube (CNT)-based NO_2 sensors have traditionally faced challenges with slow sensor signal readout and recovery times. The conventional steady-state sensing approach needs long readout times often exceeding two hours [17] and recovery at room temperature takes even longer, often over twelve hours [25].

To accelerate both the readout and recovery times while maintaining the ultra-low power, this work actively uses the self-heating-enabled recovery of suspended CNTs. Suspended CNTs, due to their minimal heat dissipation into substrates, can undergo substantial heating with just microwatts of power, functioning like an ideal heating filament [61]. This self-heating facilitates faster desorption of gas molecules from the sensor surface, with the fastest demonstrated recovery taking 10 minutes [25].

Moreover, the sensor surface resetting via self-heating presents an opportunity to reduce readout times. By monitoring the re-adsorption of gas molecules on the reset surface, the initial slope of the sensor's transient response can be utilized to quickly estimate the concentration of the analyte, a process inherently faster than waiting for a steady-state signal [16, 17].

Earlier applications of these two methods significantly reduce sensor signal recovery time to 10-45 minutes [25,76] and transient readout to 10-40 minutes [17, 76]. These developments are notable, but for practical applications, further reduction of recovery and readout times to just a few minutes is necessary. For initial slope sensing, a simple linear fitting technique was employed to yield an estimated LOD of 23 ppb.

In this chapter, an understanding of the physical phenomena during the self-heating-assisted desorption and subsequent re-adsorption is leveraged to further optimize the operational protocols of suspended CNTFET-based NO_2 sensors. The examination of the temperature changes along the self-heated nanotube and their impact on different sensing areas identifies an optimal gate bias (V_{gs}) that significantly speeds up the recovery of the device current (I_d) signal. As a result, the experiments consistently demonstrate sensor signal recovery within 1 minute. Upon the return of gas molecules to the surface of the reset nanotube after a self-heating pulse, a physics-based nonlinear response function is applied to analyze the sensor's transient behavior. A particularly interesting transient signal is the initial slope of the device current (I_d) signal. This methodology enables the determination of NO_2 concentrations within 1-5 minutes. Cyclical switching between *recovery mode* to reset the sensor and *sensing mode* to gauge NO_2 concentration, illustrated in Figure 5.1, completes a measurement cycle in just a few minutes. The peak power usage in this operation is below 6 μW, with an extrapolated LOD ranging from 9 to 61 ppb. The reproducibility of this operation is validated with additional devices, which demonstrate similar key performance metrics.

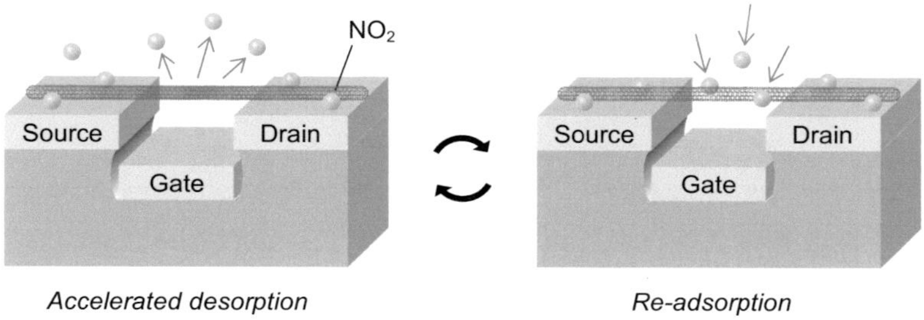

Figure 5.1: Illustrations of accelerated gas desorption from the self-heated, suspended CNT channel (left) and re-adsorption of gas molecules as the CNT cools down (right). The sensor operation method described in this chapter alternates between these two modes. Reproduced from the author's own publication [75] (open access available under the Creative Commons CC BY license).

5.2 Experimental setup and methods

The gas sensor devices used in this chapter are fabricated as described in Section 3.2 and Appendix A, with a few differences in detail. For example, the CNTs were grown at 850 °C, not 825 °C, as a result of the CNT growth parameter optimization described in Section 4.2.2. In addition, the gate-CNT distance L_g is 2 μm, not 1 μm. Details about the individual devices can be found in Appendix B.

For dry transfer of nanotubes, both manual dry transfer described in Section 3.2.5 and automated transfer described in Chapter 4 are used. Regardless of the nanotube transfer method, the contact improvement protocols described in Section 3.2.4 and 3.2.6 are applied.

The main investigation of this chapter analyzes one device (Device ID: SJDS157-D3) in depth. All the analysis procedures are described in detail using this device, and then the performance of this device will be compared with additional devices analyzed in the same way.

This main device was fabricated by manual dry transfer, and its transfer characteristic (I_d-V_{gs}) and output characteristic (I_d-V_{sd}) are presented in Figure 5.2. The output characteristic (I_d-V_{sd}) clearly shows a negative differential conductance (NDC) regime indicating significant Joule heating in the suspended CNT channel. The bias conditions in this regime are used to induce nanotube self-heating in the *recovery mode* of the gas measurements.

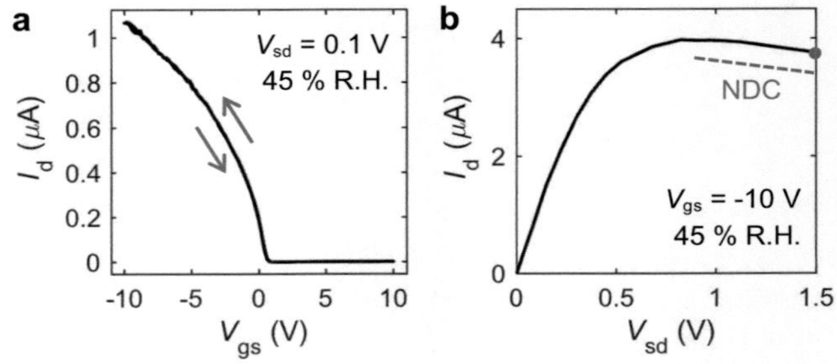

Figure 5.2: (a) Hysteresis-free transfer characteristic (I_d-V_{gs}) of a gas sensor device (Device ID: SJDS157-D3) in ambient conditions (23 C, 45 % R.H.) obtained with a pulsed V_{gs} sweep [59]. (b) Output characteristic (I_d-V_{sd}) of the device showing the peak current at V_{sd} of 0.8 V followed by a negative differential conductance (NDC) regime. The bias conditions of $V_{sd} = 1.5$ V and $V_{gs} = -10$ V (marked with circle), which belong to the NDC regime, are used to induce the nanotube self-heating effect in the gas measurements. Reproduced from the author's own publication [75] (open access available under the Creative Commons CC BY license).

In the experimental setup for gas sensing measurements, synthetic air composed of 20 % O_2 and 80 % N_2 by volume served as the diluent gas. Humidity control was achieved by channeling the gas through a deionized water-filled bubbler connected to a thermostat (Julabo F-34). The gas test chamber was equipped with temperature (Pt-1000) and humidity (Honeywell HIH-4000 Series) sensors to closely monitor the conditions, which were maintained at 23 ± 1 °C and 45 ± 1.5 % R.H. to reproduce typical operational environments. Block schematics of the measurement setup is presented in Figure 5.3. Experimental support of Cristina Gentili and Kishan Thodkar with the gas measurement setup is highly appreciated.

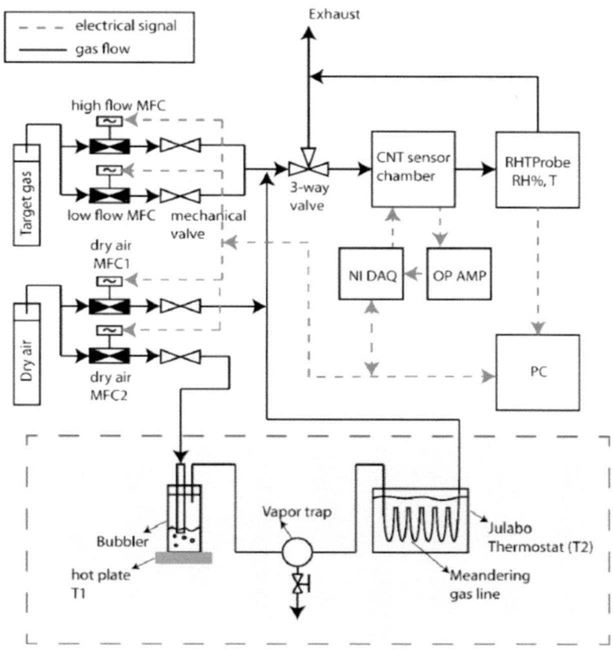

Figure 5.3: Overview of the gas sensor measurement setup used in this work. Illustration courtesy of Kishan Thodkar.

The electrical characterizations were executed with the aid of LabVIEW-controlled DAQ boards (National Instruments NI 6289 and NI BNC-2110). In recovery mode, a bias condition within the negative differential conductance (NDC) regime (V_{sd} = 1.5 V and V_{gs} = -10 V) was applied constantly for 1 minute. In sensing mode, transfer characteristics (I_d-V_{gs}) were recorded every 250 ms using a pulsed V_{gs} sweep, and a constant V_{sd} of 0.1 V. As described earlier, the measurements constantly alternated between these two modes.

5.3 Results and analysis

5.3.1 Fast sensor recovery with optimized gate bias

Previous research has shown that utilizing the self-heating effect in suspended nanotubes can enable fast sensor signal recovery within 10-45 minutes [25, 76]. This recovery speed is obtained by monitoring the ON-state current at a substantial gate bias (V_{gs} < -1.5 V for p-type devices) as the sensor signal, as commonly done for other CNTFET-based chemical sensors [155, 156].

In this section, it is demonstrated that the device current I_d at a gate bias (V_{gs}) near the threshold voltage (V_{th}) recovers faster than I_d at V_{gs} in the ON-state. Experiments presented later suggest that I_d at a gate bias (V_{gs}) near the threshold voltage (V_{th}) recovers after 1 minute of nanotube self-heating. Therefore, this work suggests that V_{gs} near V_{th} is an optimal bias setting for the sensing mode. To understand why V_{gs} near V_{th} is an optimal bias setting, it is crucial to consider the changes in different sensing regions during the nanotube self-heating phase.

Hsu *et al.* [157] used Raman G mode downshifts to demonstrate that the largest temperature change, ΔT, occurs in the middle of the tube, with a sharp decrease near the contacts, as qualitatively illustrated in Figure 5.4. The rate, at which gas molecules desorb, is expected to depend exponentially on the local temperature, T, meaning desorption is significantly faster in the middle of the tube compared to at the contacts. Consequently, the time needed for bulk desorption – i.e. from the suspended nanotube channel – is primarily governed by ΔT in the transitional areas highlighted by ovals in Figure 5.4, where temperatures reach intermediate ΔT levels. These varied sensing regions correspond to distinct responses in the device's transfer characteristics (I_d-V_{gs}). While adsorption on the CNT channel (bulk) shifts the threshold voltage (V_{th}) [18, 25], adsorption at the contacts affects the slope of the transfer characteristic [158, 159]. Thus, monitoring I_d near V_{th}, rather than in the ON-state, should enable quicker detection of bulk desorption due to nanotube self-heating.

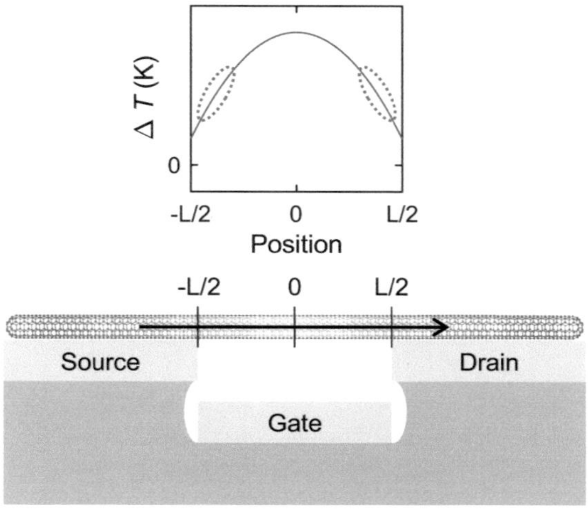

Figure 5.4: Qualitative illustration of the temperature profile along a self-heated, suspended carbon nanotube. ΔT in the transitional nanotube regions between the contacts and the middle of the nanotube (marked with ovals) has been shown to be smaller than that in the middle of the tube [157]. Reproduced from the author's own publication [75] (open access available under the Creative Commons CC BY license).

To verify this hypothesis, a gas measurement was designed with regular 1-minute self-heating pulses at increasing NO_2 concentrations. Then, I_d near V_{th} (V_{gs} = 0.5 V) was monitored in sensing mode. The results demonstrate that after 1 minute of nanotube self-heating, I_d drops to a consistent level across all NO_2 concentrations, suggesting successful recovery of the CNT channel (Figure 5.5a). However, I_d at V_{gs} = -3.0 V from the same measurement shows incomplete recovery after 1-minute self-heating pulses already at concentrations as low as 50-100 ppb. The incomplete signal recovery is more pronounced at higher concentrations of 500-1000 ppb. This can be explained by the non-uniform temperature profile, which makes desorption at the contacts not possible within just 1 minute.

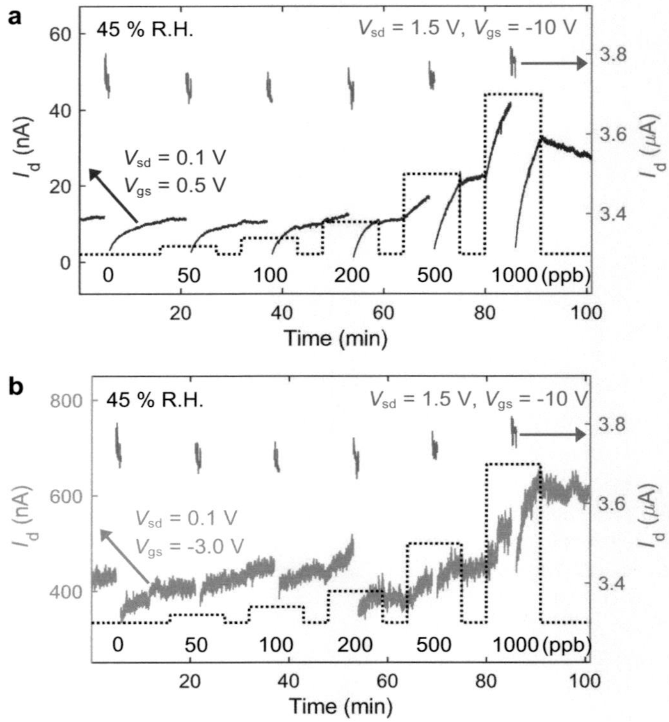

Figure 5.5: Comparison of device current (I_d) of the same device measured (a) near threshold voltage, V_{gs} = 0.5 V, and (b) at a larger gate bias, V_{gs} = -3 V. (a) and (b) are from the same gas measurement with 1-minute self-heating pulses and increasing concentrations of NO_2, which are represented by dashed black lines. (a) At V_{gs} = 0.5 V, I_d undershoots to approximately the same level after 1-minute self-heating pulses at all NO_2 concentrations. At V_{sd} = 0.1 V and V_{gs} = 0.5 V, the device is in the sensing mode, and V_{sd} = 1.5 V and V_{gs} = -10 V, the device is in the recovery mode. (b) At V_{gs} = -3 V, I_d does not recover completely after self-heating pulses. At V_{sd} = 0.1 V and V_{gs} = -3.0 V, the device is in the sensing mode, and V_{sd} = 1.5 V and V_{gs} = -10 V, the device is in the recovery mode. Reproduced from the author's own publication [75] (open access available under the Creative Commons CC BY license).

Transition state theory helps to estimate the ΔT that is required for bulk desorption within 1 minute, as it provides an equation relating desorption time (τ_{des}) to the binding energy (E_b) [160]:

$$\tau_{\text{des}} = \nu_0^{-1} e^{\frac{-E_b}{k_B T}} \quad (5.1)$$

Here, ν_0 represents the attempt frequency (approximately 10^{12} for NO_2), k_B is the Boltzmann constant (8.62×10^{-5} eV K^{-5}), and T is the temperature in Kelvin. At room temperature, the recovery of CNT-based NO_2 sensors takes approximately 12 hours [25], and this corresponds to desorption barrier E_b = -1 eV. Equation 5.1 suggests that a reduction in recovery time from 12 hours to 1 minute should require a temperature increase of about 50 K in the transitional regions (marked with ovals in Figure 5.4). Previous investigations by Hsu *et al.* showed a ΔT of approximately 40 K in similar regions at an electrical heating power of 2.67 μW. The heating power during the recovery mode of the measurement shown in Figure 5.5 is around 5.6 μW. In addition, while Hsu *et al.* write that NDC is not observed with their heating conditions, the recovery mode in Figure 5.5 clearly belongs to the NDC regime (Figure 5.2. These support the feasibility of a temperature increase close to 50 K in our device, which enables effective bulk recovery within 1 minute.

The observation here confirms that employing a gate bias V_{gs} near V_{th} during sensing mode enables the observation of noticeably faster sensor signal recovery when the self-heating effect in a suspended nanotube is used as a recovery method.

5.3.2 Fast sensor readout with nonlinear transient sensing

This section discusses how to use the sensor's re-adsorption phase after a self-heating pulse to determine the analyte concentration. As alluded to earlier, this can be done by utilizing the initial slope or rate of the transient response of the sensor current I_d. Previous work on initial slope-based sensing applies simple linear fits to sensor transients and assumes single-species adsorption dynamics [16, 17, 76]. However, consideration of a multi-species adsorption dynamics can provide a more accurate physical description.

A hypothesis here is that the self-heating of the nanotube accelerates the desorption of not only NO_2 molecules but also other environmental species, including N_2, O_2, and H_2O. The re-adsorption of environmental species makes up the *baseline* transient signal $\Delta I_{d,env}$. To characterize $\Delta I_{d,env}$, a control experiment was conducted, where 1-minute self-heating pulses were applied repeatedly under ambient conditions without NO_2. As shown in Figure 5.6a, after each pulse, I_d immediately undershoots and then returns to its original, pre-heating level. This supports the hypothesis that environmental species are desorbed through the self-heating of the nanotube and then gradually re-adsorb as the nanotube returns to ambient temperature. This kind of undershoot following self-heating was also noted in previous studies, both in environments with and without NO_2 [25]. Figure 5.6b shows five baseline transients in a single plot, demonstrating their striking reproducibility. A rising exponential, a(1-e^{-bt}), was fitted to the first minute's data points, yielding

$$\Delta I_{d,\mathrm{env}} = (11.41\,\mathrm{nA}) \cdot \left(1 - e^{-(0.23\,\mathrm{min}^{-1})t}\right) \tag{5.2}$$

Here, t is the time after heating in minutes and $\Delta I_{d,env}$ is measured in nA. The same exponential model was fitted to data for the first 3 and 5 minutes as well (see insets of Figure 5.6b). The changes of fitted parameters a and b as functions of fit length are illustrated in Figure 5.6c. The effectiveness of a single exponential fit in describing the transients suggests that the response, for up to 5 minutes after self-heating-induced recovery, is primarily governed by a single dominant environmental species.

In Figure 5.7, which shows all transient data points for the first 13 minutes (the longest duration available in the gas measurement), one can see that a single exponential fit works well up to ~10 minutes. The deviations after 10 minutes might hint at the onset of the second species at this point, but the data here is not conclusive, and a more specifically designed experiment with longer transients would be needed for more insights.

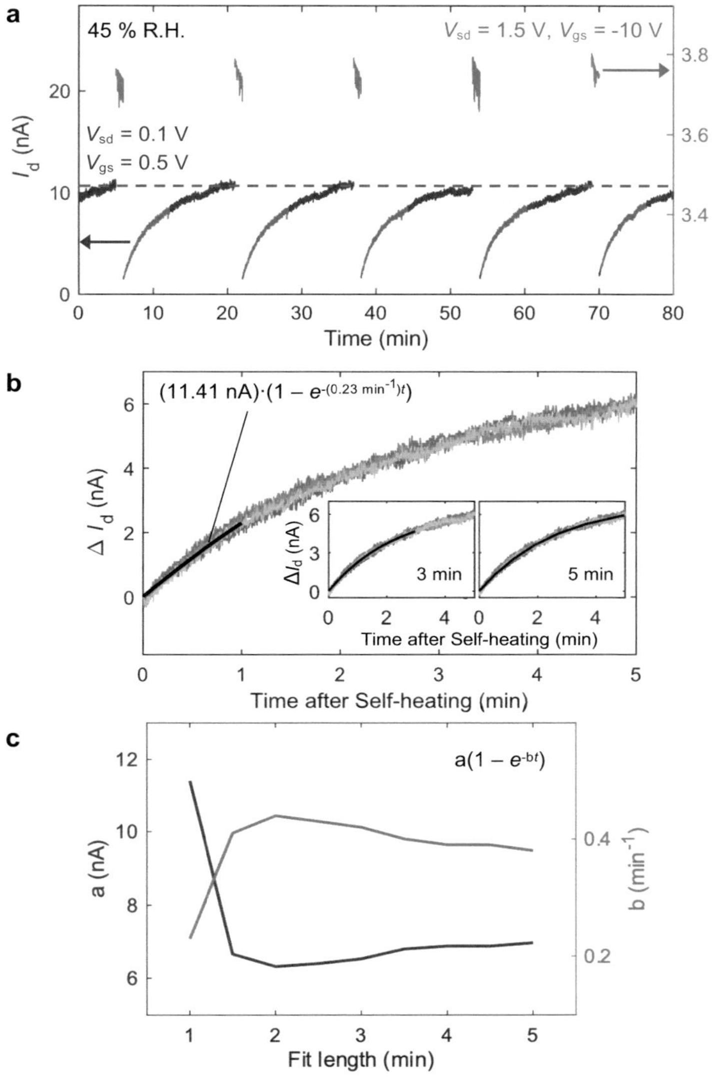

Figure 5.6: Baseline transients in ambient conditions without NO_2. (a) I_d undershoots after each self-heating pulse and then returns to the pre-heating level marked with a gray dashed line. Dark gray represents I_d in the sensing mode and light gray represents recovery mode. (b) Five I_d transients corresponding to the highlighted parts in (a) plotted together. These baseline transients show strong reproducibility. Black solid line represents an exponential fit. Insets: exponential fits to the data points in the first 3 and 5 minutes. (c) Fitted parameters a and b as functions of fit length. (a) and (b) are Reproduced from the author's own publication [75] (open access available under the Creative Commons CC BY license).

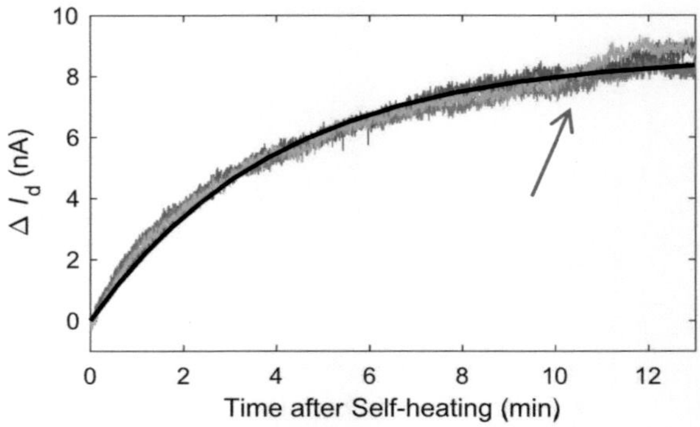

Figure 5.7: A single exponential fit to longer, 13-minute baseline transients. Deviations from the single exponential fit start to appear after 10 minutes (marked by arrow).

In subsequent experiments, NO_2 was introduced into the system and its concentration was raised incrementally. Figure 5.8a displays I_d data from two repeated measurements (the first measurement was already shown in Figure 5.5a). In the presence of multiple gas species, multiple Langmuir isotherms are expected. In the time windows directly after the self-heating pulses, the experimental environment can be considered in a simplified way to be a two-species system composed of NO_2 and the dominant environmental species outlined earlier. Under these conditions, ΔI_d can be modeled as a polynomial that includes individual responses $\Delta I_{d,env}$ and $\Delta I_{d,NO_2}$, higher-order terms such as $\Delta {I_{d,env}}^2$ and $\Delta {I_{d,NO_2}}^2$, and cross terms such as $\Delta I_{d,env} \cdot \Delta I_{d,NO_2}$. The simplest model that can be created with these terms is the sum of the individual responses:

$$\Delta I_d = \Delta I_{d,\text{env}} + \Delta I_{d,\text{NO}_2} \tag{5.3}$$

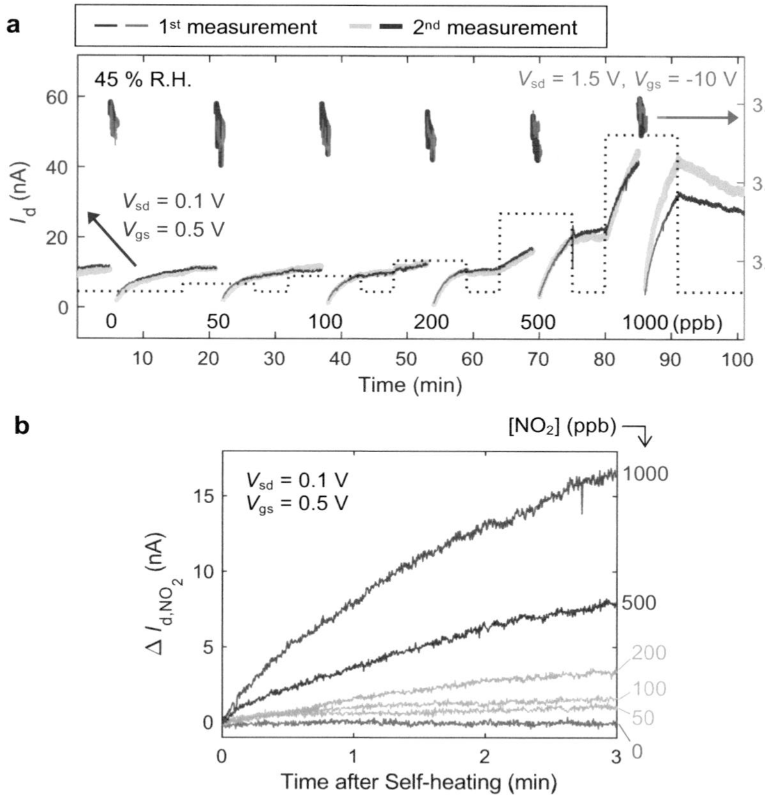

Figure 5.8: NO_2 gas measurements with repeated 1-minute self-heating pulses. (a) I_d data from two gas measurements with increasing NO_2 concentrations. (b) Transient $\Delta I_{d,NO_2}$ responses directly after self-heating pulses (corresponding to the highlighted parts of the 1st measurement in (a)). Following Equation 5.3, the baseline exponential fit found in the absence of NO_2 ($\Delta I_{d,env}$, Figure 5.6b) has been subtracted from ΔI_d data points in the sensing mode to plot $\Delta I_{d,NO_2}$. Reproduced from the author's own publication [75] (open access available under the Creative Commons CC BY license).

In Figure 5.8a, the $\Delta I_{d,NO_2}$ transient is noticeably steeper at higher NO_2 concentrations. By subtracting the baseline response obtained from the control experiment ($\Delta I_{d,env}$, Figure 5.6b) from the transients measured in the presence of NO_2 (ΔI_d), the $\Delta I_{d,NO_2}$ component can be isolated. The $\Delta I_{d,NO_2}$ waveforms plotted together in Figure 5.8b are clearly distinguishable even at the lowest NO_2 concentrations of the measurement.

The baseline subtraction procedure that yield a $\Delta I_{d,NO_2}$ is illustrated in Figure 5.9. The $\Delta I_{d,env}$ fit function is a continuous function, so at each time step, a $\Delta I_{d,env}$ value corresponding to the time step is subtracted from the ΔI_d data point.

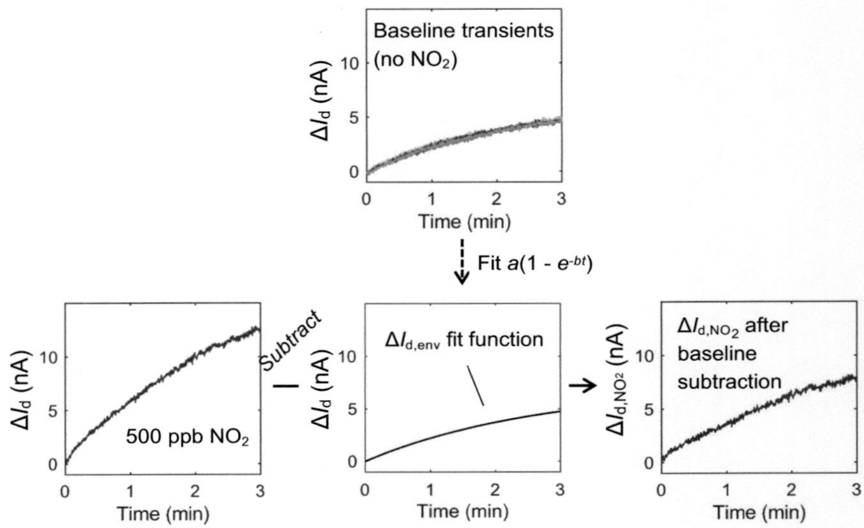

Figure 5.9: Graphical explanation of the baseline subtraction procedure using 500 ppb data as an example.

The Langmuir adsorption model is widely used to model NO_2 adsorption on carbon nanotubes [161–163]. Hence, an exponential fitting function derived from Langmuir dynamics can be used to determine the initial rate of the $\Delta I_{d,NO_2}$ transient:

$$\Delta I_{d,\mathrm{NO_2}} = \frac{\partial I}{\partial \theta}\theta(t) = \frac{\partial I}{\partial \theta}\frac{K_{\mathrm{ads}}p}{K_{\mathrm{ads}}p + K_{\mathrm{des}}}(1 - e^{-(K_{\mathrm{ads}}p + K_{\mathrm{des}})t}) \tag{5.4}$$

θ denotes the fraction of adsorption sites occupied by NO_2 (surface coverage), p is the partial pressure of NO_2, and K_{ads} and K_{des} are the adsorption and desorption rate constants, respectively. The term $\partial\theta/\partial I$ is a conversion factor between surface coverage and device current.

For this main device, we assume a constant $\partial I/\partial\theta$ of 32 nA. K_{des} is determined to be 0.026 min^{-1}, and this value is found by fitting a decaying exponential function ($a \cdot e^{-K_{des}t}$) to the unassisted desorption phase of this device, specifically from Minute 91 to Minute 101 in the measurements illustrated in Figure 5.8a. The other fit parameter, a, is 36.7 nA.

As explained earlier, the signal of interest is the initial rate of $\Delta I_{d,NO_2}$ transient immediately after the signal recovery ($d\Delta I_{d,NO_2}/dt$ at $t = 0$). This signal will be referred to as R for convenience. Equation 5.4 suggests that

$$R = \frac{\partial I}{\partial \theta} K_{\text{ads}} p \tag{5.5}$$

By using Equation 5.4 as a fit function, one can find $K_{ads}p$, which is the only unknown in the equation, and as a result, determine the initial rate R [17]. At all concentrations depicted in Figure 5.8b, Equation 5.4 is used to fit the data points within the first 1, 3, or 5 minutes to determine the initial rates. As demonstrated in Figure 5.10a with 1000 ppb data, the fit Equation 5.4 to the $\Delta I_{d,NO_2}$ transient works well regardless of the length of the fit window. Table 5.1 and 5.2 summarize the initial rate R and adsorption coefficient K_{ads} values with varying length of fit function at different NO_2 concentrations.

Table 5.1: Initial rate R with varying length of fit function at different NO_2 concentrations. Unit: nA min^{-1}.

	50 ppb	100 ppb	200 ppb	500 ppb	1000 ppb
1 min	0.45	0.85	1.46	3.34	9.60
3 min	0.41	0.67	1.38	3.49	8.58
5 min	0.33	0.51	1.04	3.11	8.40

Table 5.2: K_{ads} with varying length of fit function at different NO_2 concentrations. Unit: min^{-1} ppm^{-1}.

	50 ppb	100 ppb	200 ppb	500 ppb	1000 ppb
1 min	0.27	0.26	0.22	0.20	0.29
3 min	0.25	0.20	0.21	0.21	0.26
5 min	0.20	0.15	0.16	0.19	0.25

A sensor tested at the same temperature is expected to have consistent K_{ads} at different concentrations, and the K_{ads} values in Table 5.2 meet this expectation.

The results of this analysis implies that a simple addition of individual responses $\Delta I_{d,env}$ and $\Delta I_{d,NO_2}$ (Equation 5.3) adequately explains the net ΔI_d transients for up to 5 minutes after nanotube self-heating without involving higher-order terms and cross terms in this particular experiment.

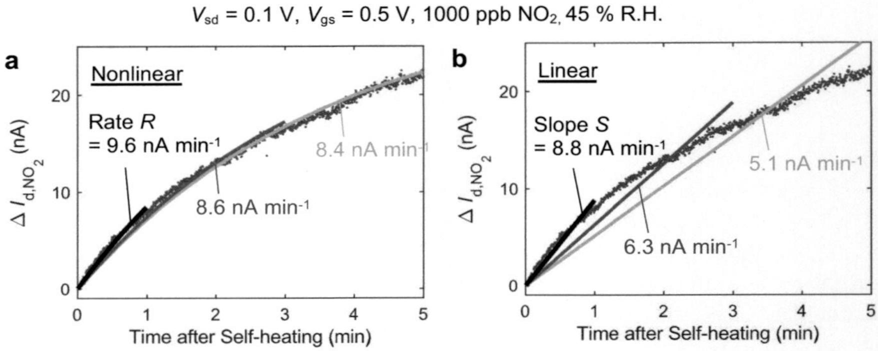

Figure 5.10: Comparison of transient sensing based on physics-based nonlinear fits and linear fits using 1000 ppb NO_2 data as an example. (a) Determination of the initial rate R of $\Delta I_{d,NO_2}$ transient via Equation 5.4 Initial rate R does not vary significantly with the readout time. (b) Determination of the initial slope S via linear fits. 1 ppm NO_2 transient. A linear model underestimates the initial slope S of the concave $\Delta I_{d,NO_2}$ transient as the readout window becomes longer. Reproduced from the author's own publication [75] (open access available under the Creative Commons CC BY license).

Equation 5.3 and Equation 5.4 offer a more accurate description of the physics than linear fit-based initial slope sensing. Now, the next interesting question is whether this method also offers any practical advantage in terms of sensor performance metrics.

For this, the Langmuir model-based nonlinear fit is compared with the conventional linear fit (Figure 5.10). It was mentioned earlier that the initial rate R determined by this method is consistent across various readout times (Figure 5.10a). In contrast, the linear fits tend to underestimate the initial slope S of the concave sensor transients as the readout window extends (Figure 5.10b).

This has an implication for the sensor's limit of detection (LOD) defined as $3\sigma_{noise}/(\text{sensitivity})$. Here, σ_{noise} represents the standard deviation of initial rates or slopes from five baseline transients depicted in Figure 5.6a after the baseline subtraction illustrated in Figure 5.9.

Consider sensitivity first. Sensitivity of the two methods evolves differently. In the nonlinear transient sensing method, the sensitivity remains relatively stable across various readout times, as shown by the slopes of the dashed lines in Figure 5.11a. In contrast, in linear transient sensing, the diminishing initial slope with longer readout times results in a decrease in sensitivity, as shown by the slopes of the dashed lines in Figure 5.11b.

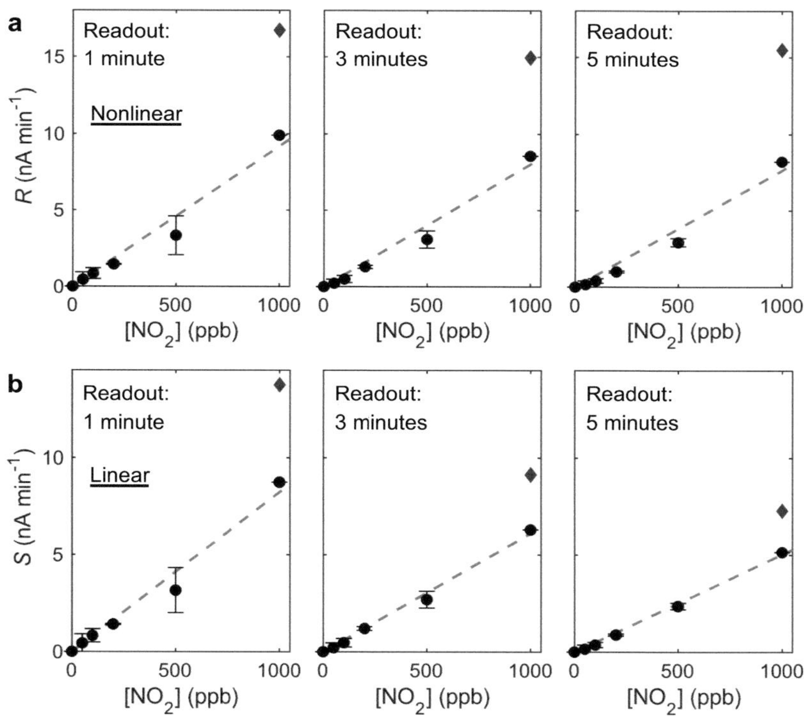

Figure 5.11: Sensor response of the (a) nonlinear transient sensing method and (b) linear transient sensing method calculated with 1-minute (left), 3-minute (middle), and 5-minute (right) readout windows. Dots represent means and error bars represent standard deviations across measurements; dashed lines are linear sensor response functions. The 1 ppm NO_2 data point from the second measurement is considered an outlier and is marked by a rhombus. Reproduced from the author's own publication [75] (open access available under the Creative Commons CC BY license).

Next, σ_{noise} should be considered. Figure 5.12a illustrates that σ_{noise} decreases with increasing readout time for both linear and nonlinear transient sensing methods. This reduction is expected as longer readout times allow more extensive averaging, which is beneficial from a statistical perspective.

The final result is that extending the readout window does not improve the LOD of the linear transient sensing method so much, although a smaller σ_{noise} with a longer readout window should actually improve the LOD. This is because of the loss in sensitivity (Figure 5.11b). As a result, the extrapolated LOD, which is calculated to be 66 ppb for a 1-minute readout, only goes down to 35 ppb for a 5-minute readout (Figure 5.12b).

On the other hand, the nonlinear transient sensing method fully benefits from longer readout windows – and lower σ_{noise} – because it maintains its sensitivity. The extrapolated LOD decreases from 61 ppb for a 1-minute readout to 9 ppb for a 5-minute readout (Figure 5.12b).

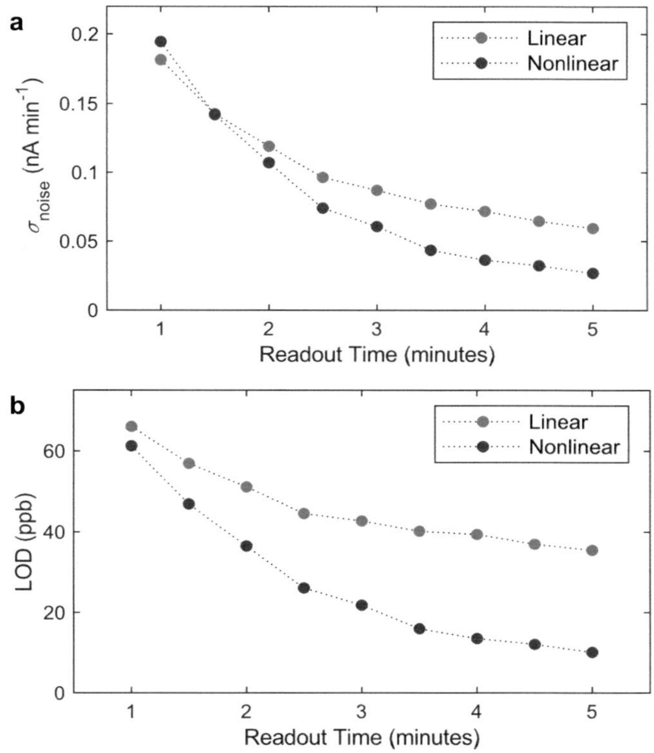

Figure 5.12: (a) σ_{noise} and (b) extrapolated limit of detection (LOD) as a function of readout time for the linear (light gray) and nonlinear (dark gray) transient sensing methods. Reproduced from the author's own publication [75] (open access available under the Creative Commons CC BY license).

5.3.3 Nanotube quality analysis with Raman spectroscopy

Following a comprehensive series of gas sensing measurements, the structural integrity of the nanotube was assessed using Raman spectroscopy. The Raman spectra were acquired using a Renishaw inVia confocal Raman microscope, focusing on the midpoint of the suspended carbon nanotube. A laser with a 515 nm wavelength was directed onto the nanotube, which was magnified 50 times by the objective lens. The scans covered the area with a power density set at 30.2 W/mm^2, taking 60 seconds per individual measurement spot.

This study was the first in the literature to test nanotube self-heating for sensor recovery in a typical field environment that includes humid air and NO_2. A key concern not addressed in earlier research was whether repeated self-heating cycles in the presence of NO_2, a strong oxidizing agent, could degrade the nanotube's structural quality. The nanotube on the device described extensively so far in this chapter underwent approximately three hours of self-heating in total. After all gas measurements, a confocal Raman microscope was used to produce a Raman image shown in Figure 5.13a showing the nanotube suspended between the electrodes. Multiple Raman spectra were collected, with one example depicted in Figure 5.13b. None of the spectra showed the D mode (1300-1400 cm^{-1}), which is typically indicative of structural defects [164]. This absence of the D mode suggests the nanotube remained structurally undamaged despite the extended period of self-heating in an NO_2-containing environment.

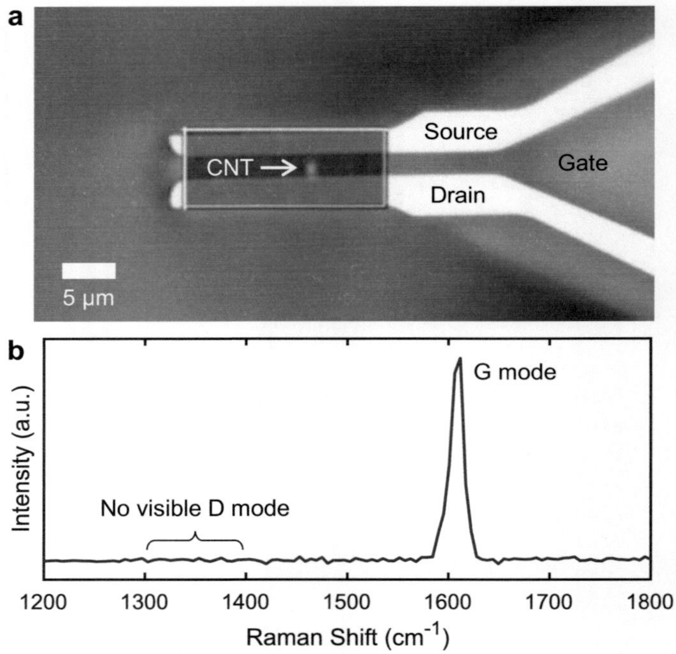

Figure 5.13: (a) Raman image of the carbon nanotube gas sensor reported in this chapter, superimposed on an optical microscopy image. A filter for carbon nanotube's characteristic G mode was applied. (b) Raman spectra from the suspended channel obtained after the gas measurements including 3 hours of self-heating time. The D mode, indicative of structural damage, was not detected. Reproduced from the author's own publication [75] (open access available under the Creative Commons CC BY license).

5.3.4 Additional devices

Manually assembled device

To validate the repeatability of the gas sensing and recovery behavior, additional devices were tested in experiments similar to the one described earlier. These devices confirmed the trends observed in the first device, exhibiting key performance metrics within a comparable range.

The second device (Device ID: SJDS160-D3) showed an ON-resistance of 132 kΩ defined at a source-drain voltage (V_{sd}) of 0.1 V and a gate bias (V_{gs}) of -10 V. The device's output characteristic (I_d-V_{sd}) showed the onset of an NDC behavior at V_{sd} = 1.2 V (V_{gs} = -10 V). Figure 5.14 presents the I_d record from a similarly designed gas measurement with increasing NO_2 concentrations. After 1-minute self-heating induced by V_{sd} = 1.5 V and V_{gs} = -10 V, I_d consistently undershot and then returned to the pre-heating level, a behavior that was observed in the first device (Figure 5.14a). The heating power during the recovery mode was approximately 4.2 μW. In the readout mode, K_{des} was determined using the data points from Minute 188 and Minute 191 of the gas measurement (Figure 5.14a), and it was found be to 0.020 min^{-1}.

The initial rate of the $\Delta I_{d,NO_2}$ transient was, again, linearly dependent on NO_2 concentration, as illustrated in Figure 5.14b-c. As this device reproduced the overall trends of the first device, its extrapolated LOD was also in a similar range. With a readout time of 5 minutes, extrapolated LOD was 23 ppb. This device shared the same geometrical dimensions with the first device and was assembled manually.

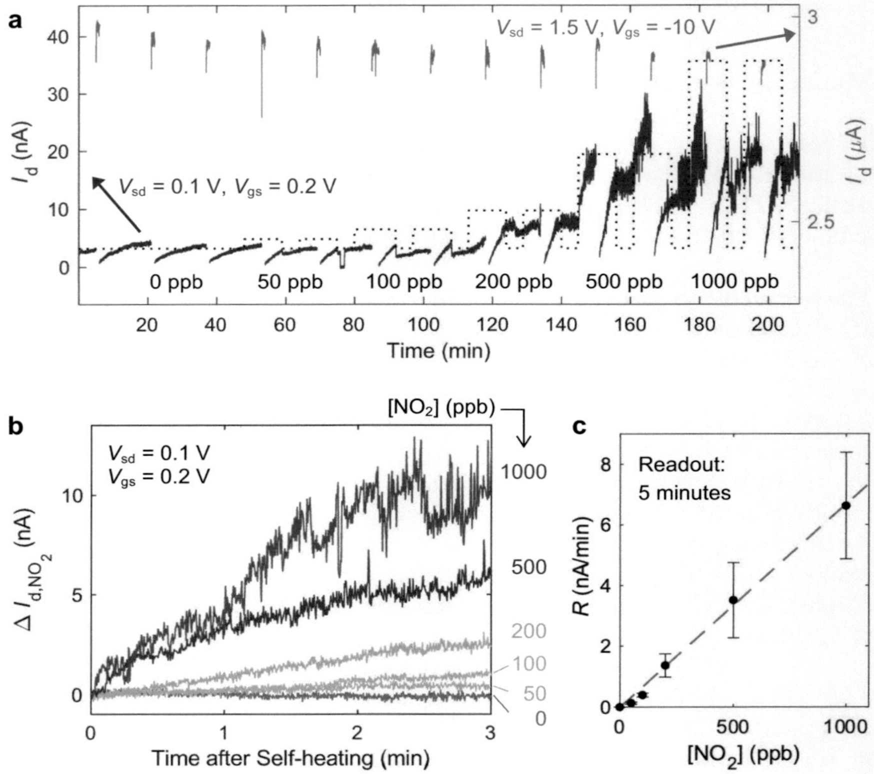

Figure 5.14: NO_2 gas measurement of the second device (Device ID: SJDS160-D3). (a) Time series of I_d data recorded in a gas measurement with increasing NO_2 concentration. (b) $\Delta I_{d,NO_2}$ transient responses after self-heating. The data points from the last sensing mode at each concentration are plotted. Baseline transient subtraction was performed as illustrated in Figure 5.9. (c) Sensor response with a readout time of 5 minutes. Dots represent means and error bars represent standard deviations across measurements; dashed lines are linear sensor response functions. Reproduced from the author's own publication [75] (open access available under the Creative Commons CC BY license).

Automatically assembled device

The third device (Device ID: DW22-C32-D4) was assembled automatically, as described in Chapter 4. This device had a smaller L_{sd} of 0.5 μm (500 nm) and L_{gate} of 0.12 μm (120 nm) (Figure 4.13). In the first two devices, the electrode patterns were written with optical lithography, as described in

Appendix A. In case of this device, e-beam lithography was used instead to obtain the small L_{sd}.

The I_d record from a gas measurement, shown in Figure 5.15 suggests that automatically assembled suspended CNTFETs perform similarly as NO_2 sensors to manually assembled devices. Again, self-heating pulses in the recovery mode consistently brought I_d back to the pre-heating level at a power of 4.8 μW (Figure 5.15a). In the readout mode, K_{des} was determined using the data points from Minute 124 and Minute 129 of the gas measurement (Figure 5.15a), and it was found be to 0.043 min^{-1}. Figure 5.15b-c show NO_2 concentration-dependent initial rate of the $\Delta I_{d,NO_2}$ transient. With a readout time of 5 minutes, the extrapolated LOD was 20 ppb.

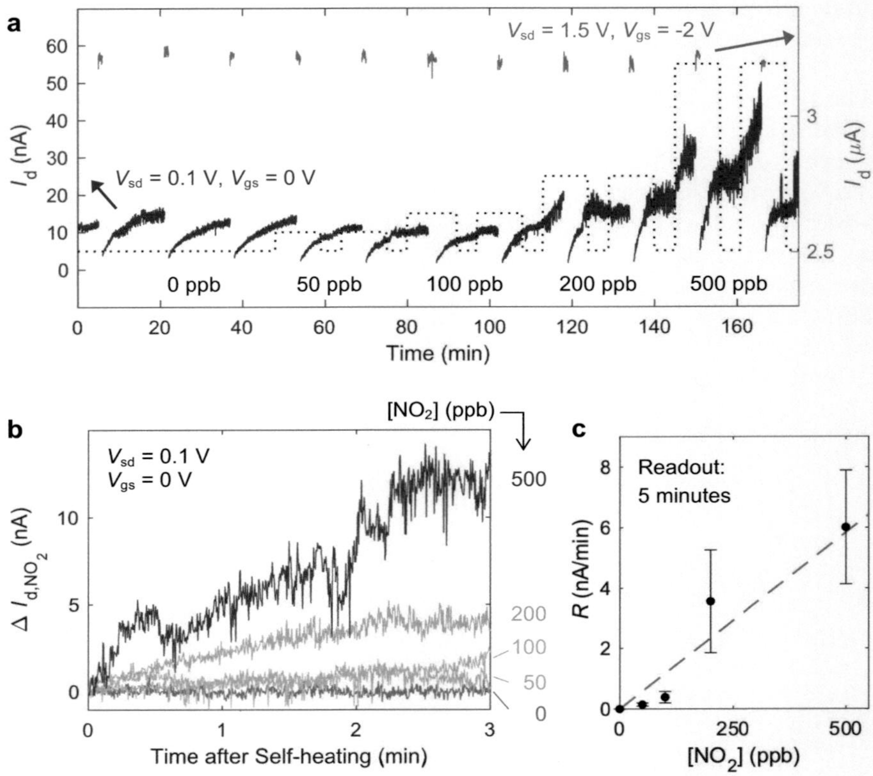

Figure 5.15: NO_2 gas measurement of the third device (Device ID: DW22-C32-D4). (a) Time series of I_d data recorded in a gas measurement with increasing NO_2 concentration. (b) $\Delta I_{d,NO_2}$ transient responses after self-heating. The data points from the last sensing mode at each concentration are plotted. Baseline transient subtraction was performed as illustrated in Figure 5.9. (c) Sensor response with a readout time of 5 minutes. Dots represent means and error bars represent standard deviations across measurements; dashed lines are linear sensor response functions.

Summary of devices

Table 5.3 summarizes the key performance parameters of the measured gas sensor devices. These parameters include K_{des}, K_{ads}, R_{on}, peak power during self-heating, and (estimated) LOD. As shown in the table, these performance metrics all lie in similar ranges whether the nanotube is transferred manually or automatically.

Table 5.3: Summary of the key performance parameters of devices used for gas measurements. R_{on} is defined at V_{sd} = 0.1 V, V_{gs} = -10 V. Extrapolated LOD is defined with a 5-minute readout window. The self-heating pulses, or recovery modes, are 1 minute long.

Device	Assembly	K_{des}	K_{ads}	R_{on}	Peak power	LOD
Device 1 (SJDS157-D3)	Manual	0.026 min^{-1}	0.19 min^{-1} ppm^{-1}	97 kΩ	5.6 μW	9 ppb
Device 2 (SJDS160-D3)	Manual	0.020 min^{-1}	0.28 min^{-1} ppm^{-1}	132 kΩ	4.2 μW	23 ppb
Device 3 (DW22-C32-D4)	Automated	0.043 min^{-1}	0.32 min^{-1} ppm^{-1}	130 kΩ	4.8 μW	20 ppb

These similar performance metrics can be attributed to a few factors. The contact improvement methods introduced in Chapter 3 are shown to effectively reduce the variability of R_{on} of CNTFETs (Figure 3.9). The nanotube assembly procedures involve Raman-based pre-selection of nanotubes, which reduces the chance of producing multiple-CNT devices. In addition, among the assembled devices, only those that show sufficiently low R_{on} and an NDC behavior are selected for gas measurements. These measures are likely to lead to a narrow subset of devices that behave similarly to one another.

5.4 Chapter summary

In summary, a fast, ultra-low-power NO2 sensor based on a self-heated, suspended CNT has been demonstrated. The suspended nanotube channel is reset by accelerating gas desorption via nanotube self-heating and the initial rate of sensor transients is read out as gas molecules re-adsorb on the nanotube surface to determine the analyte concentration. By considering the temperature changes during nanotube self-heating and their effect on different sensing regions, it is found that a gate bias (V_{gs}) near the threshold voltage (V_{th}) enables the observation of sensor signal recovery in 1 minute at a very low power of ~5.6 μW, supporting an assumption that a gate bias (V_{gs}) near the threshold voltage (V_{th}) helps to mainly observe the changes in the CNT bulk region. Additionally, readout based on a physics-based, nonlinear determination of the initial rate of the sensor transient requires only 1-5 minutes. Alternating between sensing mode and recovery mode could complete the measurement cycle in 2 minutes, yielding an extrapolated LOD of 61 ppb. Extending the readout time to a few minutes could achieve a lower LOD; an extrapolated LOD of 9 ppb has been demonstrated with a readout time of 5 minutes, marking a 4-fold improvement in LOD over previously reported linear fit-based sensing. The reader is kindly directed back to Table 2.2 if curious about how the CNT-based NO_2 sensor of this work compares to other nanostructure-based NO_2 sensors.

These performance improvements have been reproduced with additional devices – assembled both manually and automatically. The first demonstration of the compatibility of automatically assembled CNTFETs with the operation method presented in this chapter proposes potential of high-volume fabrication of fast, ultra-low-power gas sensors based on carbon nanotubes. Table 5.3 presented a summary of successfully measured gas sensor devices.

The nonlinear fit function used in this chapter (Equation 5.4) originates from the traditional Langmuir dynamics for a single species. However, given that the model (Equation 5.3) assumes the simultaneous re-adsorption of multiple species, applying a multi-species adsorption model like the extended Langmuir (EL) model to describe $\Delta I_{(d,NO_2)}$ would be more complete and allow the analysis of multicomponent gas mixtures [165]. In the Appendix D, a comparison is made between Equation 5.4 and an equation derived from the multi-species, extended Langmuir (EL) model. This comparison demonstrates that Equation 5.4 serves as a suitable approximation for the first few minutes after nanotube self-heating.

One important question to ask at the end of this chapter is how the nonlinear transient readout scheme would work in real-world settings, where NO_2 concentrations do not change in stepwise increments but continuously. The readout scheme remains applicable unless there is a significant change in the analyte concentration within the brief, 1- to 5-minute time window needed for the readout. Data from air quality monitoring stations in urban and traffic areas indicate that NO_2 concentrations usually change slowly, typically just a few ppb per hour [15]. Nonetheless, should there be a sudden change in concentration within the readout window, the adsorption dynamics would become more complex, and the algorithm would yield a concentration value that is intermediate between the lowest and highest NO_2 concentrations observed during that interval.

The improvements reported in this work advance practical NO_2 sensing with carbon nanotubes for low-power mobile or Internet-of-Things (IoT) sensing systems. Future work is anticipated to build upon these studies and further explore other practically important topics such as empirical LOD determined with repeated ppb-level measurements, cross-sensitivity analysis, influence of varying humidity levels, and drift effects due to contact degradation.

6 Conclusions and Outlook

6.1 Conclusions

This thesis presents an investigation into the fabrication of suspended carbon nanotube (CNT) devices and their operation as fast, ultra-low-power NO_2 sensors, focusing on key aspects such as nanotube-electrode contact, high-throughput device fabrication, and fast sensor operation. The key findings from each investigation are summarized below, supplemented by critical evaluations of the studies. Outlook on future research directions closes the chapter.

Contact Improvement

The experiments in Chapter 3 demonstrated a significant correlation between the presence of oxygen-containing adsorbates on surfaces of Pd bottom contacts and the high contact resistance at the CNT-electrode interface. It was shown that electrode etching with Ar-ion immediately prior to nanotube transfer effectively removes these adsorbed species, resetting the surface. This electrode resetting markedly improved CNT-electrode contact, which was further enhanced by post-transfer thermal annealing. The combined pre- and post-transfer treatments resulted in a substantial reduction in the median ON-resistance (R_{on}) of suspended CNTFETs by an order of magnitude, with the interquartile range reduced by more than two orders of magnitude. This improvement in contact resistance was crucial for the self-heating operation of suspended CNT gas sensors, which was further investigated in the later part of the thesis.

This study focused on demonstrating the improvement in device resistance with a rather limited set of treatment parameters. Therefore, a comparative study with several groups of devices undergoing different Ar-ion cleaning parameters and thermal annealing parameters could further optimize the contact improvement effect. A recent study by Althuon *et al.* shows a good example of refined optimization [100]. For example, in addition to adopting the Ar-ion electrode cleaning and thermal annealing proposed by this work, they add a current annealing step to the protocols. Furthermore, they investigate the correlation between thermal annealing power and contact improvement effect. As a result, they report a good number of devices with resistance below 50 kΩ, which was not demonstrated in this work.

High-throughput Device Fabrication with Robotic Assembly

Chapter 4 addressed a new process flow for high-throughput fabrication of suspended CNTFETs. This process flow was designed around automated, robotic assembly of carbon nanotubes. The first step in the process was the batch growth of suspended CNTs. Through a systematic multi-parameter optimization process, the optimal growth conditions were identified, significantly increasing the number of transferrable, suspended CNTs on the growth substrate. The implementation of large-scale growth substrates with thousands of trenches further increased the number of suspended CNTs grown in each CVD growth run. Despite some non-uniformity in growth activity, the batch growth on large-scale substrates yielded approximately 24 % single-CNT trenches.

The automated assembly of CNTs was realized using a nanoassembly machine, achieving a transfer rate of approximately 46 transfers per hour with a 70 % yield of electrically active devices. The results demonstrated the efficacy of the automated robotic assembly for high-throughput fabrication of ultraclean CNTFETs.

Although this chapter provides the first statistics on R_{on}, V_{th}, and SS of automatically assembled CNT devices, a much larger dataset is necessary to obtain a more comprehensive overview. Future studies should aim to analyze over 1,000 or 10,000 devices to provide truly meaningful statistics. Additionally, there is a need for detailed analysis of variations in device characteristics such as R_{on}, V_{th}, and SS, along with investigations into the root causes of these variations.

Furthermore, before scalable fabrication can be claimed, it is essential to characterize and presumably improve the stability of devices under operational conditions. This study primarily focused on the yield of electrically active devices; however, for the fabrication technology to be truly relevant, a more rigorous definition of yield is needed. This definition should consider the aforementioned parameters or the specific use case of the fabricated devices.

While this work reports the automated assembly of CNTs into devices, the parameters of the assembly motions were not thoroughly studied. The transfer rate used in this study (~50 transfers/hour) is significantly lower than the nanoassembly system's current performance limit, which is approximately ~400 transfers/hour. This means that future studies have plenty of room to investigate the effect of assembly parameters – such as transfer speed and forces during the transfer – on CNTFET performance with the current nanoassembly system. Investigations beyond the limit of the current system (>400 transfers/hour) can consider changes in system software, working mechanism, and design principles.

Fast, Ultra-low-lower NO_2 Sensors

A fast, ultra-low-power NO_2 sensor based on a self-heated, suspended CNT was demonstrated in Chapter 5. By accelerating gas desorption via nanotube self-heating and utilizing a physics-based, nonlinear determination of the initial rate of sensor transients, the sensor achieved signal recovery in 1 minute and fast readout in 1-5 minutes at a peak power of ~5.6 μW. The measurement cycle, alternating between readout and recovery modes, could be completed in 2 minutes with an extrapolated limit of detection (LOD) of 61 ppb. Extending the readout time to 5 minutes further improved the LOD to 9 ppb. These performance improvements were reproduced with additional devices, both manually and automatically assembled. The compatibility of automatically assembled CNTFETs with the sensor operation method underscores the potential for high-volume fabrication of fast, ultra-low-power gas sensors.

However, the nonlinear transient sensing method introduced in this work relies significantly on the assumptions of a highly controlled measurement environment. It is unclear whether the scheme would be as effective in a more complex measurement environment or in a field test.

In addition, the gas measurements in this work used a gas concentration profile that increases monotonously. The operation scheme should be verified with a decreasing gas concentration profile as well. A randomly changing gas concentration would be even better.

Although this work reproducibly demonstrates the proposed operation method with three devices, a much larger sample size in the range of a few hundres is needed to obtain statistics about the gas sensor performance. When this much data becomes available, meaningful correlations between FET performance metrics and gas sensor performance metrics are likely to be found.

6.2 Outlook

The advancements reported in this thesis pave the way for further research and development in the field of CNT-based electronic devices and sensors. Future work could focus on the following areas:

Tailored Assembly of Carbon Nanotubes

This work demonstrated the robotic assembly of pre-selected carbon nanotubes, but the potential for pre-selection extends beyond the current implementation. Future research can focus on enhancing the selectivity of the assembly process, not only by pre-selecting CNTs based on the CNT location and count but also by their specific properties such as electronic type (semiconducting, metallic, or small-gap semiconducting), diameter, and defect density. This level of selectivity can enable the fabrication of highly specialized devices tailored for specific applications, improving performance and uniformity in device characteristics. Advanced characterization techniques, combined with machine learning algorithms, can be employed to rapidly and accurately classify CNTs before assembly, further refining the precision of the automated nanoassembly process.

Long-Term Stability of Gas Sensors

Investigating the long-term stability and reliability of the CNT gas sensors is crucial, especially with a focus on the effects of contact degradation, drift, and environmental factors on sensor performance. While this thesis presented some preliminary data on sensor stability, the timeframe used in the study is far shorter than what the sensors would need to endure in real-life applications. Future research can extend the stability tests to longer periods and test the devices in field tests or simulated real-life conditions to assess how the sensors perform over months or years.

Selectivity of Gas Sensors

The gas measurements presented in this thesis were conducted under controlled laboratory conditions with a limited number of gas species. While these controlled tests provide insights into the sensor's performance and capabilities, real-world environments present a more complex challenge. Future research can focus on enhancing the selectivity of the CNT-based gas sensors to accurately distinguish between different gas species in the presence of multiple potential interferents. Possible options include the use of sensor arrays and additional functionalization techniques to improve the sensor's specificity. Such improvements will be critical for deploying CNT-based sensors in practical applications, such as air quality monitoring and industrial safety, in which the ability to reliably detect specific target gases in a complex mixture is essential.

By addressing these areas, future research can build upon the work presented in this thesis, further enhancing the practical applicability and scalability of CNT-based devices not only in NO_2 sensing but also in a wide range of technological applications.

A Appendix A - Process Run Sheets

A.1 Growth substrates

	Process	Tool [Location]	Parameters
1	**WAFER PREPARATION**		
1.1	SOI wafer (400 µm Si handle – 2 µm SiO_2 – 5 µm Si device)		
2	**PHOTOLITHOGRAPHY 1 (FORK STRUCTURES ON DEVICE LAYER)**		
2.1	Mask cleaning	OptiWet ST30 [BRNC]	Recipe: 5 inch Mask clean 2 x PIR, chamber rinse cold
2.2	Wafer cleaning	Wet bench [BRNC]	Acetone dip/ultrasonic bath: 2 min IPA dip: 1 min O_2 plasma ashing: 600 W, 5 min
2.3	HMDS	Wet bench [BRNC]	
2.4	Photoresist coating	Wet bench [BRNC]	Resist: AZ1512 Time: 40 sec Spin speed: 4000 RPM Ramp speed: 2000 RPM Post-bake: 110 °C, 5 min
2.5	Exposure	Aligner MA6 [BRNC]	
2.6	Development	Wet bench [BRNC]	Recipe: AZ1512
2.7	Si device layer dry etch	DRIE [BRNC]	Recipe: (to be added) Etch time: <5 min Target etch depth: 6 µm
2.8	Stripping with Piranha	Wet bench [BRNC]	Acetone rinsing: 1 min Piranha dip: 120 °C, 5 min Water rinse: 5 min
2.9	Protective SiO_2 on device layer	PECVD [BRNC]	Recipe: 000 Std SiO2 – Dave-like lower SiH4 flow Deposition time: 20 min

3	**PHOTOLITHOGRAPHY 2 (HANDLE LAYER)**		
3.1	Mask cleaning	OptiWet ST30 [BRNC]	Recipe: 5 inch Mask clean 2 x PIR, chamber rinse cold
3.2	Wafer cleaning	Wet bench [BRNC]	Acetone dip/ultrasonic bath: 2 min IPA dip: 1 min O_2 plasma ashing: 600 W, 5 min
3.3	Protective photoresist on device layer	Wet bench [BRNC]	Resist: AZ4562 Time: 40 sec Spin speed: 4000 RPM Ramp speed: 2000 RPM Post-bake: 110 °C, 7 min
3.4	HMDS	Wet bench [BRNC]	
3.5	Photoresist coating	Wet bench [BRNC]	Resist: AZ4562 Time: 40 sec Spin speed: 4000 RPM Ramp speed: 2000 RPM Post-bake: 110 °C, 7 min
3.6	Exposure with backside alignment	Aligner MA8 [BRNC]	
3.7	Development	Wet bench [BRNC]	Recipe: AZ4562
3.8	Si handle layer dry etch	DRIE [BRNC]	Attach carrier wafer on device layer Recipe: (ASE recipe, to be updated) Etch time: 40-45 min Target etch depth: 400 µm
3.9	Carrier wafer detachment	Wet bench [BRNC]	DMSO dip: 120 °C, 15-20 min
3.10	Stripping with Piranha	Wet bench [BRNC]	Acetone rinsing: 1 min Piranha dip: 120 °C, 5 min Water rinse: 5 min
4	**OXIDE LAYER ETCH**		
4.1	Buried SiO_2 etch	HF bench [BRNC]	BHF 1:7 Time: 30 min
5	**THERMAL OXIDATION**		
5.1	Cleaning with Piranha	Wet bench [BRNC]	Acetone rinsing: 1 min Piranha dip: 120 °C, 5 min Water rinse: 5 min
5.2	SiO_2 wet thermal oxidation	Furnace [BRNC] (ask BRNC OpTeam)	Temperature: 1200 °C Target thickness: 100 nm

A.2 Device substrates

	Process	Tool [Location]	Parameters
1	**WAFER PREPARATION**		
1.1	SiO_2 wet thermal growth on 500-μm Si wafer	Furnace [BRNC] (ask BRNC OpTeam)	Temperature: 1200 °C Target thickness: 200 nm - 2 μm
1.2	Al_2O_3 deposition	ALD thermal Picosun [BRNC]	Recipe: AL2O3SEB Temperature: 300 °C Cycles: 300 Target thickness: 30 nm TMA pulse time: 0.1 sec TMA purge time: 4 sec H_2O pulse time: 0.1 sec H_2O purge time: 0.1 sec
1.3	Wafer cleaning	Wet bench [BRNC]	Acetone dip/ultrasonic bath: 2 min IPA dip: 1 min O_2 plasma ashing: 600 W, 5 min
1.4	HMDS on topside	Wet bench [BRNC]	
1.5	Protective photoresist on topside	Wet bench [BRNC]	Resist: AZ4533 Time: 40 sec Spin speed: 4000 RPM Ramp speed: 2000 RPM Post-bake: 110 °C, 5 min
1.6	Backside SiO_2/Al_2O_3 removal	HF bench [BRNC]	BHF 1:7 Time: 30 min
2	**PHOTOLITHOGRAPHY 1 (PATTERNING OF S/D ELECTRODES)**		
2.1	Mask cleaning	OptiWet ST30 [BRNC]	Recipe: 5 inch Mask clean 2 x PIR, chamber rinse cold
2.2	Wafer cleaning/topside protective photoresist removal	Wet bench [BRNC]	Acetone dip/ultrasonic bath: 2 min IPA dip: 1 min O_2 plasma ashing: 600 W, 5 min
2.3	HMDS	Wet bench [BRNC]	
2.4	Photoresist coating	Wet bench [BRNC]	Resist: AZ4533 Time: 40 sec Spin speed: 4000 RPM Ramp speed: 2000 RPM Post-bake: 110 °C, 5 min
2.5	Exposure	Aligner MA6 [BRNC]	
2.6	Development	Wet bench [BRNC]	Recipe: AZ4533
2.7	Al_2O_3 dry etch	ICP 2 [BRNC]	Recipe: 000 Al2O3 etch BCl3/Cl2 Feb 2020 Etch time: 2 min 30 sec
2.8	SiO_2 dry etch	RIE 1 [BRNC]	Recipe: 000 SiO2 Ar CHF3 100W 16 nm/min Target etch depth: SiO_2 thickness
2.9	Stripping	Wet bench [BRNC]	Acetone dip/ultrasonic bath: 2 min IPA dip: 1 min DMSO dip: 120 °C, 10 min Water rinse: 2 min O_2 plasma ashing: 600 W, 5 min

3	**PHOTOLITHOGRAPHY 2 (PATTERNING OF MESAS)**		
3.1	Mask cleaning	OptiWet ST30 [BRNC]	Recipe: 5 inch Mask clean 2 x PIR, chamber rinse cold
3.2	HMDS	Wet bench [BRNC]	
3.3	Photoresist coating	Wet bench [BRNC]	Resist: AZ4533 Time: 40 sec Spin speed: 4000 RPM Ramp speed: 2000 RPM Post-bake: 110 °C, 5 min
3.4	Exposure	Aligner MA6 [BRNC]	
3.5	Development	Wet bench [BRNC]	Recipe: AZ4533
3.6	Al_2O_3 dry etch	ICP 2 [BRNC]	Recipe: 000 Al2O3 etch BCl3/Cl2 Feb 2020 Etch time: 2 min 30 sec
3.7	SiO_2 dry etch	RIE 1 [BRNC]	Recipe: 000 SiO2 Ar CHF3 100W 16 nm/min Target etch depth: SiO_2 thickness
3.8	Si dry etch	DRIE Si Etcher [BRNC]	Recipe: 0_ASE_10C Cycles: 30 Target thickness: 30 µm
3.9	Stripping	Wet bench [BRNC]	Acetone dip/ultrasonic bath: 2 min IPA dip: 1 min DMSO dip: 120 °C, 10 min Water rinse: 2 min O_2 plasma ashing: 600 W, 5 min
4	**UNDERCUT ETCH**		
4.1	Pre-bake	Wet bench [BRNC]	Hot plate: 200 °C, 5 min
4.2	HF vapor etch	HF vapor etch [BRNC]	Recipe: 2 Time: 240 sec
5	**ELECTRODE FORMATION**		
5.1	Metal evaporation	Evaporator [BRNC]	Layer 1: Cr, 1 nm Layer 2: Pd, 40 nm

6	**PARTIAL DICING**		
6.1	Adhesion promoter (HMDS)	HMDS hot plate [BRNC]	Recipe: 110C HMDS
6.2	Photoresist coating	Wet bench [BRNC]	Resist: AZ4533 Time: 40 sec Spin speed: 4000 RPM Ramp speed: 2000 RPM Post-bake: 110 °C, 5 min
6.3	Dicing	Dicer [BRNC] (Submit wafer with dicing job request)	Target dicing depth: 325 µm

B Appendix B - List of Devices

Table B.1: List of suspended CNTFETs used for gas measurements and other experimental investigations.

Device ID	$\mathbf{L}_{sd}$	$\mathbf{L}_{gate}$	**Presented in**
SJDS83-D3	2.8 μm	1 μm	Figure 3.14a-b
SJDS85-D3	2.8 μm	1 μm	Figure 3.14c-d
SJDS97-D1	2.8 μm	1 μm	Figure 3.15a
SJDS95-D1	2.8 μm	1 μm	Figure 3.15b
SJDS95-D2	2.8 μm	1 μm	Figure 3.15b
SJDS53-D2	2.8 μm	1 μm	Figure 3.16
SJDS52-D3	2.8 μm	1 μm	Figure 3.16
SJDS157-D3	2 μm	2 μm	Figure 5.2, 5.5, 5.6, 5.7, 5.8, 5.9, 5.10, 5.11, 5.12, 5.13
SJDS160-D3	2 μm	2 μm	Figure 5.14
DW22-C32-D4	0.5 μm	0.12 μm	Figure 5.15

C Appendix C - Batch CVD Growth Runs

In Section 4.2.4, three CVD runs for batch growth of suspended CNTs were discussed. The result of the first CVD run was presented in Figure 4.8 and Table 4.3 and 4.4. This Appendix presents the results of the second and third CVD runs.

Table C.1: Counts of single-CNT trenches in the second CVD run.

Substrate	**Top**	**Left**	**Right**
0° Fork	21	13	30
90° Fork	25	18	34
180° Fork	23	21	28
270° Fork	20	20	26
Average	**23.2**	**Standard Deviation**	**5.5**

Table C.2: Counts of multiple-CNT trenches in the second CVD run.

Substrate	**Top**	**Left**	**Right**
0° Fork	16	11	20
90° Fork	33	47	26
180° Fork	5	62	29
270° Fork	18	37	18
Average	**26.8**	**Standard Deviation**	**15.4**

Table C.3: Counts of single-CNT trenches in the third CVD run.

Substrate	**Top**	**Left**	**Right**
0° Fork	16	11	21
90° Fork	31	27	23
180° Fork	32	41	19
270° Fork	19	6	19
Average	**22.1**	**Standard Deviation**	**9.2**

Table C.4: Counts of multiple-CNT trenches in the third CVD run.

Substrate	**Top**	**Left**	**Right**
0° Fork	14	8	11
90° Fork	19	18	8
180° Fork	24	24	5
270° Fork	18	0	10
Average	**13.2**	**Standard Deviation**	**7.2**

D Appendix D - Langmuir Adsorption Model

The content of this Appendix is largely reported in [75].

Chapter 5 uses the conventional Langmuir adsorption model to analyze the adsorption dynamics of NO_2. A limitation of the conventional Langmuir adsorption model is that it only considers one adsorbate. However, the experimental findings in Chapter 5 suggest that nanotube self-heating induces desorption and re-adsorption of other environmental species as well. Therefore, a multi-species adsorption model, such as the extended Langmuir (EL) model, can provide a more complete description of the NO_2 adsorption dynamics [165]. In the EL model, the isotherm for species i is:

$$\theta_{\infty,i}(p) = \frac{K_i p_i}{1 + \sum_{j=1}^{N} K_j p_j} \tag{D.1}$$

Here, $K_i = K_{\text{ads},i}/K_{\text{des},i}$. Chapter 5 suggests that, during the re-adsorption phase immediately following self-heating-assisted recovery, two primary species can be considered: (1) NO_2 and (2) one dominant environmental species. NO_2 can be denoted with subscript $i = 1$, and the dominant environmental species with $i = 2$. In the EL model with two species, the dynamics of the surface coverage of NO_2, $\theta_1(t)$, would be expressed as follows:

$$\theta_1(t) = \frac{K_1 p_1}{1 + K_1 p_1 + K_2 p_2} \left(1 - e^{-K_{\text{des},1}(1+K_1 p_1 + K_2 p_2)(1-\theta_{0,2})t}\right) \tag{D.2}$$

Here, $\theta_{0,2}$ represents the initial surface coverage of species $i = 2$ immediately after self-heating. The initial rate of Equation D.2 is:

$$\left.\frac{d\theta_1(t)}{dt}\right|_{t=0} = (1-\theta_{0,2})K_{\text{ads},1}p_1 \tag{D.3}$$

By comparing Equation D.3 and Equation 5.5, one sees that the initial rate of a single-species model, Equation 5.5, differs from that of a two-component model, Equation D.3, by a factor of $(1-\theta_{0,2})$. If one used Equation D.3 to determine the sensor signal R in Chapter 5, R would be scaled by $(1-\theta_{0,2})$ at all NO_2 concentrations.

Figure 5.8 in Chapter 5 shows current undershoots following self-heating pulses, and these undershoots were attributed to the desorption of environmental species. The reproducibility of the undershoots suggests that $\theta_{0,2}$ is almost constant after 1-minute self-heating pulses. Therefore, using the single-species Langmuir model to determine the initial rate of $\Delta I_{d,\text{NO2}}$ would consistently reflect the scaling factor $(1-\theta_{0,2})$ at all NO_2 concentrations and would not affect the overall trend of the linear sensor response (Figure 5.11) in the measurement environment used in this work.

Bibliography

[1] World Health Organization, "Ambient air pollution: A global assessment of exposure and burden of disease," World Health Organization, Tech. Rep., 2016.

[2] S. Nedelcu, "Energy Efficient Analog Mixed-Signal Front Ends for CNT-FET NO2 Air-Quality Nanosensors," Ph.D. dissertation, ETH Zurich, 2022, DOI: http://doi.org/10.3929/ethz-b-000587135.

[3] European Commission, "Directive 2008/50/EC of the European Parliament and of the Council of 21 May 2008 on ambient air quality and cleaner air for Europe," *Official Journal of the European Union*, vol. 152, pp. 1–44, 2008.

[4] K. Annamalai and I. K. Puri, *Combustion Science and Engineering*. CRC Press, 2006, DOI: http://doi.org/10.1201/9781420003789.

[5] A. Helwig, G. Müller, G. Sberveglieri, and M. Eickhoff, "On the low-temperature response of semiconductor gas sensors," *J. Sensors*, vol. 2009, pp. 1–17, 2009, DOI: http://doi.org/10.1155/2009/620720.

[6] M. Hossain, J. Saffell, and R. Baron, "Differentiating NO2 and O3 at Low Cost Air Quality Amperometric Gas Sensors," *ACS Sensors*, vol. 1, no. 12, pp. 1291–1294, 2016, DOI: http://doi.org/10.1021/acssensors.6b00603.

[7] M. I. Mead, O. A. M. Popoola, G. B. Stewart, P. Landshoff, M. Calleja, M. Hayes, J. J. Baldovi, M. W. McLeod, T. F. Hodgson, J. Dicks, A. Lewis, J. Cohen, R. Baron, J. R. Saffell, and R. L. Jones, "The use of electrochemical sensors for monitoring urban air quality in low-cost, high-density networks," *Atmos. Environ.*, vol. 70, pp. 186–203, 2013, DOI: http://doi.org/10.1016/j.atmosenv.2012.11.060.

[8] J. D. Prades, R. Jimenez-Diaz, F. Hernandez-Ramirez, S. Barth, A. Cirera, A. Romano-Rodriguez, S. Mathur, and J. R. Morante, "Ultralow power consumption gas sensors based on self-heated individual nanowires," *Applied Physics Letters*, vol. 93, no. 12, 2008, DOI: http://doi.org/10.1063/1.2988265.

[9] N. D. Chinh, N. Van Toan, V. Van Quang, N. Van Duy, N. D. Hoa, and N. Van Hieu, "Comparative NO2 gas-sensing performance of the self-heated individual, multiple and networked SnO2 nanowire sensors fabricated by a simple process," *Sensors and Actuators B: Chemical*, vol. 201, pp. 7–12, 2014, DOI: http://doi.org/10.1016/j.snb.2014.04.095.

[10] O. Casals, N. Markiewicz, C. Fabrega, I. GraÌcia, C. CaneÌ, H. S. Wasisto, A. Waag, and J. D. Prades, "A Parts Per Billion (ppb) Sensor for NO2 with Microwatt Power Requirements Based on Micro Light Plates," *ACS Sensors*, vol. 4, no. 4, pp. 822–826, 2019, DOI: http://doi.org/10.1021/acssensors.9b00150.

[11] Y. H. Kim, S. J. Kim, Y.-J. Kim, Y.-S. Shim, S. Y. Kim, B. H. Hong, and H. W. Jang, "Self-activated transparent all-graphene gas sensor with endurance to humidity and mechanical bending," *ACS Nano*, vol. 9, no. 10, pp. 10 453–10 460, 2015, DOI: http://doi.org/10.1021/acsnano.5b04680.

[12] X. Yan, Y. Wu, R. Li, C. Shi, R. Moro, Y. Ma, and L. Ma, "High-Performance UV-Assisted NO2 Sensor Based on Chemical Vapor Deposition Graphene at Room Temperature," *ACS Omega*, vol. 4, pp. 14 179–14 187, 2019, DOI: http://doi.org/10.1021/acsomega.9b00935.

[13] A. Azizi, M. Dogan, H. Long, J. D. Cain, K. Lee, R. Eskandari, A. Varieschi, E. C. Glazer, M. L. Cohen, and A. Zettl, "High-Performance Atomically-Thin Room-Temperature NO2 Sensor," *Nano Letters*, vol. 20, pp. 6120–6127, 2020, DOI: http://doi.org/10.1021/acs.nanolett.0c02221.

[14] D. W. Gardner, X. Gao, H. M. Fahad, A.-T. Yang, S. He, A. Javey, C. Carraro, and R. Maboudian, "Scalable Ultra Low-Power Chemical Sensing with Metal-Organic Frameworks," in *2019 20th International Conference on Solid-state Sensors, Actuators, and Microsystems (TRANSDUCERS) â Eurosensors XXXIII*, 2019, pp. 1215–1218, DOI: http://doi.org/10.1109/transducers.2019.8808818.

[15] R. Kurtenbach, J. Kleffmann, A. Niedojadlo, and P. Wiesen, "Primary NO2 emissions and their impact on air quality in traffic environments in Germany," *Environ. Sci. Europe*, vol. 24, p. 21, 2012, DOI: http://doi.org/10.1186/2190-4715-24-21.

[16] R. Muller and E. Lange, "Multidimensional sensor for gas analysis," *Sens. Actuator*, vol. 9, p. 39â48, 1986, DOI: http://doi.org/10.1016/0250-6874(86)80005-1.

[17] P. F. Satterthwaite, S. Eberle, S. Nedelcu, C. Roman, and C. Hierold, "Transient and steady-state readout of nanowire gas sensors in the presence of low-frequency noise," *Sens. Actuators B: Chem.*, vol. 297, p. 126674, 2019, DOI: http://doi.org/10.1016/j.snb.2019.126674.

[18] J. Kong, N. R. Franklin, C. Zhou, M. G. Chapline, S. Peng, K. Cho, and H. Dai, "Nanotube molecular wires as chemical sensors," *Science*, vol. 287, no. 5453, pp. 622–625, 2000, DOI: http://doi.org/10.1126/science.287.5453.622.

[19] K. Chikkadi, *Wafer-scale integration of single-walled carbon nanotubes for gas sensing.* ETH Zurich, 2014, vol. 19, DOI: http://doi.org/10.3929/ethz-a-010154350.

[20] S. Eberle, "Ultra-clean suspended carbon nanotube gas sensors-concept for large scale fabrication and sensor characterization," Ph.D. dissertation, ETH Zurich, 2019, DOI: http://doi.org/10.3929/ethz-b-000391248.

[21] W. Kim, A. Javey, O. Vermesh, Q. Wang, Y. Li, and H. Dai, "Hysteresis caused by water molecules in carbon nanotube field-effect transistors," *Nano Letters*, vol. 3, no. 2, pp. 193–198, 2003.

[22] Y.-M. Lin, J. C. Tsang, M. Freitag, and P. Avouris, "Impact of oxide substrate on electrical and optical properties of carbon nanotube devices," *Nanotechnology*, vol. 18, no. 29, p. 295202, 2007, DOI: http://doi.org/10.1088/0957-4484/18/29/295202.

[23] M. Muoth, T. Helbling, L. Durrer, S.-W. Lee, C. Roman, and C. Hierold, "Hysteresis-free operation of suspended carbon nanotube transistors," *Nature Nanotechnology*, vol. 5, no. 8, p. 589, 2010, DOI: http://doi.org/10.1038/nnano.2010.129.

[24] K. Chikkadi, M. Muoth, W. Liu, V. Maiwald, and C. Hierold, "Enhanced signal-to-noise ratio in pristine, suspended carbon nanotube gas sensors," *Sensors and Actuators B: Chemical*, vol. 196, pp. 682–690, 2014, DOI: http://doi.org/10.1016/j.snb.2014.02.058.

[25] K. Chikkadi, M. Muoth, V. Maiwald, C. Roman, and C. Hierold, "Ultra-low power operation of self-heated, suspended carbon nanotube gas sensors," *Applied Physics Letters*, vol. 103, no. 22, p. 223109, 2013, DOI: http://doi.org/10.1063/1.4836415.

[26] A. Jungen, S. Hofmann, J. C. Meyer, C. Stampfer, S. Roth, J. Robertson, and C. Hierold, "Synthesis of individual single-walled carbon nanotube bridges controlled by support micromachining," *Journal of Micromechanics and Microengineering*, vol. 17, no. 3, p. 603, 2007, DOI: http://doi.org/10.1088/0960-1317/17/3/024.

[27] R. Tormo-Queralt, C. Møller, D. Czaplewski, G. Gruber, M. Cagetti, S. Forstner, N. Urgell-Ollé, J. Sanchez-Naranjo, C. Samanta, C. Miller, and A. Bachtold, "Novel nanotube multiquantum dot devices," *Nano Letters*, vol. 22, no. 21, pp. 8541–8549, 2022, DOI: http://doi.org/10.1021/acs.nanolett.2c03034.

[28] S. Rechnitz, T. Tabachnik, M. Shlafman, S. Shlafman, and Y. E. Yaish, "Mode coupling bi-stability and spectral broadening in buckled carbon nanotube mechanical resonators," *Nature Communications*, vol. 13, no. 1, p. 5900, 2022, DOI: http://doi.org/10.5281/zenodo.7051298.

[29] M. Muoth and C. Hierold, "Transfer of carbon nanotubes onto microactuators for hysteresis-free transistors at low thermal budget," in *2012 IEEE 25th International Conference on Micro Electro Mechanical Systems (MEMS)*. IEEE, 2012, pp. 1352–1355, DOI: http://doi.org/10.1109/MEMSYS.2012.6170417.

[30] C. C. Wu, C. H. Liu, and Z. Zhong, "One-step direct transfer of pristine single-walled carbon nanotubes for functional nanoelectronics," *Nano Letters*, vol. 10, no. 3, pp. 1032–1036, 2010, DOI: http://doi.org/10.1021/nl904260k.

[31] D.-W. You, Y.-S. Seon, Y. Jang, J. Bang, J.-S. Oh, and K.-W. Jung, "A portable gas chromatograph for real-time monitoring of aromatic volatile organic compounds in air samples," *Journal of Chromatography A*, vol. 1625, p. 461267, 2020, DOI: http://doi.org/10.1016/j.chroma.2020.461267.

[32] N. Suchatanugal, P. Chedthanorrakul, I. Thaveesangsakulthai, T. Nhujak, and C. Kulsing, "Thin layer chromatography based extraction approaches for improved analysis of volatile compounds with gas chromatography–mass spectrometry and direct analysis with gas analyzer," *Journal of Separation Science*, vol. 44, no. 2, pp. 666–675, 2021, DOI: http://doi.org/10.1002/jssc.202000917.

[33] T. Chen, X. Qi, M. Chen, B. Chen *et al.*, "Gas chromatography-ion mobility spectrometry detection of odor fingerprint as markers of rapeseed oil refined grade," *Journal of Analytical Methods in Chemistry*, vol. 2019, 2019, DOI: http://doi.org/10.1155/2019/3163204.

[34] Alphasense, "NO2-B43F," https://ametekcdn.azureedge.net/mediafiles/project/oneweb/oneweb/alphasense/products/datasheets/alphasense_no2-b43f_datasheet_en_1.pdf?revision:95cdfc22-824f-482a-b26d-d7f2f4688ea8.

[35] Bosch Sensortec, "BME680," https://www.bosch-sensortec.com/media/boschsensortec/downloads/datasheets/bst-bme680-ds001.pdf.

[36] ams OSRAM, "CCS811," https://www.mouser.com/datasheet/2/588/CCS811_DS000459_7-00-1594304.pdf.

[37] Sensirion AG, "SGP41," https://www.sensirion.com/media/documents/5FE8673C/61E96F50/Sensirion_Gas_Sensors_Datasheet_SGP41.pdf.

[38] ScioSense, "ENS160," https://eu.mouser.com/datasheet/2/1081/ENS160_Datasheet-3395902.pdf.

[39] N. Yamazoe and K. Shimanoe, "Roles of shape and size of component crystals in semiconductor gas sensors: I. response to oxygen," *Journal of the Electrochemical Society*, vol. 155, no. 4, p. J85, 2008, DOI: http://doi.org/10.1149/1.2832655.

[40] G. Korotcenkov, "Current trends in nanomaterials for metal oxide-based conductometric gas sensors: Advantages and limitations. part 1: 1d and 2d nanostructures," *Nanomaterials*, vol. 10, no. 7, p. 1392, 2020, DOI: http://doi.org/10.3390/nano10071392.

[41] R. Artzi-Gerlitz, K. D. Benkstein, D. L. Lahr, J. L. Hertz, C. B. Montgomery, J. E. Bonevich, S. Semancik, and M. J. Tarlov, "Fabrication and gas sensing performance of parallel assemblies of metal oxide nanotubes supported by porous aluminum oxide membranes," *Sensors and Actuators B: Chemical*, vol. 136, no. 1, pp. 257–264, 2009, DOI: http://doi.org/10.1016/j.snb.2008.10.056.

[42] C. Jiang, G. Zhang, Y. Wu, L. Li, and K. Shi, "Facile synthesis of SnO2 nanocrystalline tubes by electrospinning and their fast response and high sensitivity to NOx at room temperature," *CrystEngComm*, vol. 14, no. 8, pp. 2739–2747, 2012, DOI: http://doi.org/10.1039/c2ce06405g.

[43] Z. L. Wang, "Zinc oxide nanostructures: growth, properties and applications," *Journal of physics: condensed matter*, vol. 16, no. 25, p. R829, 2004, DOI: http://doi.org/10.1088/0953-8984/16/25/R01.

[44] Y. Chen, P. Xu, T. Xu, D. Zheng, and X. Li, "ZnO-nanowire size effect induced ultra-high sensing response to ppb-level H2S," *Sensors and Actuators B: Chemical*, vol. 240, pp. 264–272, 2017, DOI: http://doi.org/10.1016/j.snb.2016.08.120.

[45] M. Cuscunà, A. Convertino, E. Zampetti, A. Macagnano, A. Pecora, G. Fortunato, L. Felisari, G. Nicotra, C. Spinella, and F. Martelli, "On-chip fabrication of ultrasensitive NO2 sensors based on silicon nanowires," *Applied Physics Letters*, vol. 101, no. 10, 2012, DOI: http://doi.org/10.1063/1.4748099.

[46] M. C. McAlpine, H. Ahmad, D. Wang, and J. R. Heath, "Highly ordered nanowire arrays on plastic substrates for ultrasensitive flexible chemical sensors," *Nature Materials*, vol. 6, no. 5, pp. 379–384, 2007, DOI: http://doi.org/10.1038/nmat1891.

[47] D. Zhang, Z. Liu, C. Li, T. Tang, X. Liu, S. Han, B. Lei, and C. Zhou, "Detection of NO2 down to ppb levels using individual and multiple In2O3 nanowire devices," *Nano Letters*, vol. 4, no. 10, pp. 1919–1924, 2004, DOI: http://doi.org/10.1021/nl0489283.

[48] S. Iijima, "Helical microtubules of graphitic carbon," *Nature*, vol. 354, no. 6348, pp. 56–58, 1991, DOI: http://doi.org/10.1038/354056a0.

[49] J. Li, Y. Lu, Q. Ye, M. Cinke, J. Han, and M. Meyyappan, "Carbon nanotube sensors for gas and organic vapor detection," *Nano Letters*, vol. 3, no. 7, pp. 929–933, 2003, DOI: http://doi.org/10.1021/nl034220x.

[50] S. Ammu, V. Dua, S. R. Agnihotra, S. P. Surwade, A. Phulgirkar, S. Patel, and S. K. Manohar, "Flexible, all-organic chemiresistor for detecting chemically aggressive vapors," *Journal of the American Chemical Society*, vol. 134, no. 10, pp. 4553–4556, 2012, DOI: http://doi.org/10.1021/ja300420t.

[51] K. Xu, C. Wu, X. Tian, J. Liu, M. Li, Y. Zhang, and Z. Dong, "Single-walled carbon nanotube-based gas sensors for NO2 detection," *Integrated Ferroelectrics*, vol. 135, no. 1, pp. 132–137, 2012, DOI: http://doi.org/10.1080/10584587.2012.685424.

[52] M. Jeon, B. Choi, J. Yoon, D. M. Kim, D. H. Kim, I. Park, and S.-J. Choi, "Enhanced sensing of gas molecules by a 99.9% semiconducting carbon nanotube-based field-effect transistor sensor," *Applied Physics Letters*, vol. 111, no. 2, 2017, DOI: http://doi.org/10.1063/1.4991970.

[53] C. Cantalini, L. Valentini, L. Lozzi, I. Armentano, J. M. Kenny, and S. Santucci, "NO2 gas sensitivity of carbon nanotubes obtained by plasma enhanced chemical vapor deposition," *Sensors and Actuators B: Chemical*, vol. 93, no. 1-3, pp. 333–337, 2003, DOI: http://doi.org/10.1016/S0925-4005(03)00224-7.

[54] Y. W. Chang, J. S. Oh, S. H. Yoo, H. H. Choi, and K.-H. Yoo, "Electrically refreshable carbon-nanotube-based gas sensors," *Nanotechnology*, vol. 18, no. 43, p. 435504, 2007, DOI: http://doi.org/10.1088/0957-4484/18/43/435504.

[55] G. Chen, T. M. Paronyan, E. M. Pigos, G. U. Sumanasekera, and A. R. Harutyunyan, "The performance volatility of carbon nanotube-based devices: Impact of ambient oxygen," *Applied Physics Letters*, vol. 95, no. 12, 2009, DOI: http://doi.org/10.1063/1.3236779.

[56] G. Chen, T. M. Paronyan, E. M. Pigos, and A. R. Harutyunyan, "Enhanced gas sensing in pristine carbon nanotubes under continuous ultraviolet light illumination," *Scientific reports*, vol. 2, no. 1, p. 343, 2012, DOI: http://doi.org/10.1038/srep00343.

[57] L. Sacco, S. Forel, I. Florea, and C.-S. Cojocaru, "Ultra-sensitive NO2 gas sensors based on single-wall carbon nanotube field effect transistors: Monitoring from ppm to ppb level," *Carbon*, vol. 157, pp. 631–639, 2020, DOI: http://doi.org/10.1016/j.carbon.2019.10.073.

[58] M. Mattmann, T. Helbling, L. Durrer, C. Roman, C. Hierold, R. Pohle, and M. Fleischer, "Sub-ppm NO2 detection by Al2O3 contact passivated carbon nanotube field effect transistors," *Applied Physics Letters*, vol. 94, no. 18, 2009, DOI: http://doi.org/10.1063/1.3125259.

[59] M. Mattmann, C. Roman, T. Helbling, D. Bechstein, L. Durrer, R. Pohle, M. Fleischer, and C. Hierold, "Pulsed gate sweep strategies for hysteresis reduction in carbon nanotube transistors for low concentration NO2 gas detection," *Nanotechnology*, vol. 21, p. 185501, 2010, DOI: http://doi.org/10.1088/0957-4484/21/18/185501.

[60] K. Chikkadi, M. Muoth, and C. Hierold, "Hysteresis-free, suspended pristine carbon nanotube gas sensors," in *2013 Transducers & Eurosensors XXVII: The 17th International Conference on Solid-State Sensors, Actuators and Microsystems (TRANSDUCERS & EUROSENSORS XXVII)*. IEEE, 2013, pp. 1637–1640, DOI: http://doi.org/10.1109/Transducers.2013.6627098.

[61] E. Pop, D. Mann, J. Cao, Q. Wang, K. Goodson, and H. Dai, "Negative differential conductance and hot phonons in suspended nanotube molecular wires," *Physical Review Letters*, vol. 95, no. 15, p. 155505, 2005, DOI: http://doi.org/10.1103/PhysRevLett.95.155505.

[62] C. Fàbrega, O. Casals, F. Hernández-Ramírez, and J. Prades, "A review on efficient self-heating in nanowire sensors: Prospects for very-low power devices," *Sensors and Actuators B: Chemical*, vol. 256, pp. 797–811, 2018, DOI: http://doi.org/10.1016/j.snb.2017.10.003.

[63] A. Salehi, "A highly sensitive self heated SnO2 carbon monoxide sensor," *Sensors and Actuators B: Chemical*, vol. 96, no. 1-2, pp. 88–93, 2003, DOI: http://doi.org/10.1016/S0925-4005(03)00490-8.

[64] T. Kawano, H. C. Chiamori, M. Suter, Q. Zhou, B. D. Sosnowchik, and L. Lin, "An electrothermal carbon nanotube gas sensor," *Nano Letters*, vol. 7, no. 12, pp. 3686–3690, 2007, DOI: http://doi.org/10.1021/nl071964s.

[65] E. Strelcov, S. Dmitriev, B. Button, J. Cothren, V. Sysoev, and A. Kolmakov, "Evidence of the self-heating effect on surface reactivity and gas sensing of metal oxide nanowire chemiresistors," *Nanotechnology*, vol. 19, no. 35, p. 355502, 2008, DOI: http://doi.org/10.1088/0957-4484/19/35/355502.

[66] E. Strelcov, V. V. Sysoev, S. Dmitriev, J. Cothren, and A. Kolmakov, "Self-heated nanowire sensors: Opportunities, optimization and limitations," in *AIP Conference Proceedings*, vol. 1137, no. 1. American Institute of Physics, 2009, pp. 9–11, DOI: http://doi.org/10.1063/1.3156636.

[67] J. D. Prades, R. Jimenez-Diaz, F. Hernandez-Ramirez, S. Barth, J. Pan, A. Cirera, A. Romano-Rodriguez, S. Mathur, and J. Morante, "An experimental method to estimate the temperature of individual nanowires," *International Journal of Nanotechnology*, vol. 6, no. 9, pp. 860–869, 2009, DOI: http://doi.org/10.1504/IJNT.2009.026745.

[68] J. Seo, Y. Lim, and H. Shin, "Self-heating hydrogen gas sensor based on an array of single suspended carbon nanowires functionalized with palladium nanoparticles," *Sensors and Actuators B: Chemical*, vol. 247, pp. 564–572, 2017, DOI: http://doi.org/10.1016/j.snb.2017.03.038.

[69] J. Yun, J.-H. Ahn, D.-I. Moon, Y.-K. Choi, and I. Park, "Joule-heated and suspended silicon nanowire based sensor for low-power and stable hydrogen detection," *ACS Applied Materials & Interfaces*, vol. 11, no. 45, pp. 42 349–42 357, 2019, DOI: http://doi.org/10.1021/acsami.9b15111.

[70] C. Prajapati and N. Bhat, "Self-heating oxidized suspended pt nanowire for high performance hydrogen sensor," *Sensors and Actuators B: Chemical*, vol. 260, pp. 236–242, 2018, DOI: http://doi.org/10.1016/j.snb.2017.12.180.

[71] Y. Lim, S. Kim, Y. M. Kwon, J. M. Baik, and H. Shin, "A highly sensitive gas-sensing platform based on a metal-oxide nanowire forest grown on a suspended carbon nanowire fabricated at a wafer level," *Sensors and Actuators B: Chemical*, vol. 260, pp. 55–62, 2018, DOI: http://doi.org/10.1016/j.snb.2017.12.167.

[72] Qomaruddin, O. Casals, H. S. Wasisto, A. Waag, J. D. Prades, and C. Fàbrega, "Visible-light-driven room temperature NO2 gas sensor based on localized surface plasmon resonance: the case of gold nanoparticle decorated zinc oxide nanorods (ZnO NRs)," *Chemosensors*, vol. 10, no. 1, p. 28, 2022, DOI: http://doi.org/10.3390/chemosensors10010028.

[73] F. Schedin, A. K. Geim, S. V. Morozov, E. W. Hill, P. Blake, M. I. Katsnelson, and K. S. Novoselov, "Detection of individual gas molecules adsorbed on graphene," *Nature materials*, vol. 6, no. 9, pp. 652–655, 2007, DOI: http://doi.org/10.1038/nmat1967.

[74] G. Lu, L. E. Ocola, and J. Chen, "Reduced graphene oxide for room-temperature gas sensors," *Nanotechnology*, vol. 20, no. 44, p. 445502, 2009, DOI: http://doi.org/10.1088/0957-4484/20/44/445502.

[75] S. Jung, C. Roman, and C. Hierold, "Fast nitrogen dioxide sensing with ultralow-power nanotube gas sensors," *Advanced Sensor Research*, vol. 3, no. 1, p. 2300081, 2024, DOI: http://doi.org/10.1002/adsr.202300081.

[76] S. Nedelcu, K. Thodkar, and C. Hierold, "A customizable, low-power, wireless, embedded sensing platform for resistive nanoscale sensors," *Microsystems & Nanoengineering*, vol. 8, no. 1, p. 10, 2022, DOI: http://doi.org/10.1038/s41378-021-00343-1.

[77] N. Ramgir, N. Datta, M. Kaur, S. Kailasaganapathi, A. K. Debnath, D. Aswal, and S. Gupta, "Metal oxide nanowires for chemiresistive gas sensors: issues, challenges and prospects," *Colloids and Surfaces A: Physicochemical and Engineering Aspects*, vol. 439, pp. 101–116, 2013, DOI: http://doi.org/10.1016/j.colsurfa.2013.02.029.

[78] A. Javey, Nam, R. S. Friedman, H. Yan, and C. M. Lieber, "Layer-by-layer assembly of nanowires for three-dimensional, multifunctional electronics," *Nano Letters*, vol. 7, no. 3, pp. 773–777, 2007, DOI: http://doi.org/10.1021/nl063056l.

[79] Y. H. Ko, S. Kim, and J. S. Yu, "Drop-cast and dye-sensitized ZnO nanorod-based visible-light photodetectors," *physica status solidi (RRL)–Rapid Research Letters*, vol. 7, no. 9, pp. 659–663, 2013, DOI: http://doi.org/10.1002/pssr.201307160.

[80] S. Raychaudhuri, S. A. Dayeh, D. Wang, and E. T. Yu, "Precise semiconductor nanowire placement through dielectrophoresis," *Nano Letters*, vol. 9, no. 6, pp. 2260–2266, 2009, DOI: http://doi.org/10.1021/nl900423g.

[81] Y. Huang, X. Duan, Q. Wei, and C. M. Lieber, "Directed assembly of one-dimensional nanostructures into functional networks," *Science*, vol. 291, no. 5504, pp. 630–633, 2001, DOI: http://doi.org/10.1126/science.291.5504.630.

[82] X. Liu, Y.-Z. Long, L. Liao, X. Duan, and Z. Fan, "Large-scale integration of semiconductor nanowires for high-performance flexible electronics," *ACS Nano*, vol. 6, no. 3, pp. 1888–1900, 2012, DOI: http://doi.org/10.1021/nn204848r.

[83] H. Park, A. Afzali, S.-J. Han, G. S. Tulevski, A. D. Franklin, J. Tersoff, J. B. Hannon, and W. Haensch, "High-density integration of carbon nanotubes via chemical self-assembly," *Nature Nanotechnology*, vol. 7, no. 12, pp. 787–791, 2012, DOI: http://doi.org/10.1038/nnano.2012.189.

[84] N. Wei, P. Laiho, A. T. Khan, A. Hussain, A. Lyuleeva, S. Ahmed, Q. Zhang, Y. Liao, Y. Tian, E.-X. Ding *et al.*, "Fast and ultraclean approach for measuring the transport properties of carbon nanotubes," *Advanced Functional Materials*, p. 1907150, 2019, DOI: http://doi.org/10.1002/adfm.201907150.

[85] J. Kong, H. T. Soh, A. M. Cassell, C. F. Quate, and H. Dai, "Synthesis of individual single-walled carbon nanotubes on patterned silicon wafers," *Nature*, vol. 395, no. 6705, pp. 878–881, 1998, DOI: http://doi.org/10.1038/27632.

[86] K. Chikkadi, C. Roman, and C. Hierold, "Process control monitors for single-walled carbon nanotube based sensor fabrication processes," in *2013 IEEE 26th International Conference on Micro Electro Mechanical Systems (MEMS)*. IEEE, 2013, pp. 275–278, DOI: http://doi.org/10.1109/MEMSYS.2013.6474231.

[87] A. Vijayaraghavan, S. Blatt, D. Weissenberger, M. Oron-Carl, F. Hennrich, D. Gerthsen, H. Hahn, and R. Krupke, "Ultra-large-scale directed assembly of single-walled carbon nanotube devices," *Nano Letters*, vol. 7, no. 6, pp. 1556–1560, 2007, DOI: http://doi.org/10.1021/nl0703727.

[88] L. Jia, J. Wu, Y. Zhang, Y. Qu, B. Jia, Z. Chen, and D. J. Moss, "Fabrication technologies for the on-chip integration of 2d materials," *Small Methods*, vol. 6, no. 3, p. 2101435, 2022, DOI: http://doi.org/10.1002/smtd.202101435.

[89] E. M. Freer, O. Grachev, X. Duan, S. Martin, and D. P. Stumbo, "High-yield self-limiting single-nanowire assembly with dielectrophoresis," *Nature Nanotechnology*, vol. 5, no. 7, pp. 525–530, 2010, DOI: http://doi.org/10.1038/nnano.2010.106.

[90] M. Y. Swinkels, A. Campo, D. Vakulov, W. Kim, L. Gagliano, S. E. Steinvall, H. Detz, M. De Luca, A. Lugstein, E. Bakkers *et al.*, "Measuring the optical absorption of single nanowires," *Physical Review Applied*, vol. 14, no. 2, p. 024045, 2020, DOI: http://doi.org/10.1103/PhysRevApplied.14.024045.

[91] A. Sivan, B. Abad, T. Albrigi, O. Arif, J. Trautvetter, A. Ruiz Caridad, C. Arya, V. Zannier, L. Sorba, R. Rurali *et al.*, "Gaas/gap superlattice nanowires for tailoring phononic properties at the nanoscale: Implications for thermal engineering," *ACS applied nano materials*, vol. 6, no. 19, pp. 18 602–18 613, 2023, DOI: http://doi.org/10.1021/acsanm.3c04245.

[92] D. Vakulov, S. Gireesan, M. Y. Swinkels, R. Chavez, T. Vogelaar, P. Torres, A. Campo, M. De Luca, M. A. Verheijen, S. Koelling, L. Gagliano, J. E. M. Haverkort, F. X. Alvarez, P. A. Bobbert, I. Zardo, and E. P. A. M. Bakkers, "Ballistic phonons in ultrathin nanowires," *Nano Letters*, vol. 20, no. 4, pp. 2703–2709, 2020, DOI: http://doi.org/doi.org/10.1021/acs.nanolett.0c00320.

[93] U. E. Ali, H. Yang, V. Khayrudinov, G. Modi, Z. Cheng, R. Agarwal, H. Lipsanen, and H. Bhaskaran, "A universal pick-and-place assembly for nanowires," *Small*, vol. 18, no. 38, p. 2201968, 2022, DOI: http://doi.org/10.1002/smll.202201968.

[94] N. R. Wilson and J. V. Macpherson, "Carbon nanotube tips for atomic force microscopy," *Nature Nanotechnology*, vol. 4, no. 8, pp. 483–491, 2009, DOI: http://doi.org/10.1038/nnano.2009.154.

[95] T. Cubaynes, L. Contamin, M. Dartiailh, M. Desjardins, A. Cottet, M. Delbecq, and T. Kontos, "Nanoassembly technique of carbon nanotubes for hybrid circuit-qed," *Applied Physics Letters*, vol. 117, no. 11, 2020, DOI: http://doi.org/10.1063/5.0021838.

[96] A. Jorio and R. Saito, "Raman spectroscopy for carbon nanotube applications," *Journal of Applied Physics*, vol. 129, no. 2, 2021, DOI: http://doi.org/10.1063/5.0030809.

[97] M. Vollmann, C. Roman, M. Haluska, and C. Hierold, "Quantified stress relaxation in carbon nanotube resonators," in *2023 IEEE 36th International Conference on Micro Electro Mechanical Systems (MEMS)*. IEEE, 2023, pp. 925–928, DOI: http://doi.org/10.1109/MEMS49605.2023.10052187.

[98] Y. Wen, N. Ares, F. Schupp, T. Pei, G. Briggs, and E. Laird, "A coherent nanomechanical oscillator driven by single-electron tunnelling," *Nature Physics*, vol. 16, no. 1, pp. 75–82, 2020, DOI: http://doi.org/10.1038/s41567-019-0683-5.

[99] C. Samanta, S. De Bonis, C. Møller, R. Tormo-Queralt, W. Yang, C. Urgell, B. Stamenic, B. Thibeault, Y. Jin, D. Czaplewski, F. Pistolesi, and A. Bachtold, "Nonlinear nanomechanical resonators approaching the quantum ground state," *Nature Physics*, vol. 19, no. 9, pp. 1340–1344, 2023, DOI: http://doi.org/10.1038/s41567-023-02065-9.

[100] T. Althuon, T. Cubaynes, A. Auer, C. Sürgers, and W. Wernsdorfer, "Nano-assembled open quantum dot nanotube devices," *Communications Materials*, vol. 5, no. 1, p. 5, 2024, DOI: http://doi.org/10.1038/s43246-023-00439-3.

[101] S. Blien, P. Steger, N. Hüttner, R. Graaf, and A. K. Hüttel, "Quantum capacitance mediated carbon nanotube optomechanics," *Nature Communications*, vol. 11, no. 1, p. 1636, 2020, DOI: http://doi.org/10.1038/s41467-020-15433-3.

[102] J. Waissman, M. Honig, S. Pecker, A. Benyamini, A. Hamo, and S. Ilani, "Realization of pristine and locally tunable one-dimensional electron systems in carbon nanotubes," *Nature Nanotechnology*, vol. 8, no. 8, p. 569, 2013, DOI: http://doi.org/10.1038/nnano.2013.143.

[103] W. Dong, Z. Dai, L. Liu, and Z. Zhang, "Toward clean 2d materials and devices: Recent progress in transfer and cleaning methods," *Advanced Materials*, p. 2303014, 2023, DOI: http://doi.org/10.1002/adma.202303014.

[104] S. Masubuchi, M. Morimoto, S. Morikawa, M. Onodera, Y. Asakawa, K. Watanabe, T. Taniguchi, and T. Machida, "Autonomous robotic searching and assembly of two-dimensional crystals to build van der waals superlattices," *Nature Communications*, vol. 9, no. 1, p. 1413, 2018, DOI: http://doi.org/10.1038/s41467-018-03723-w.

[105] A. J. Mannix, A. Ye, S. H. Sung, A. Ray, F. Mujid, C. Park, M. Lee, J.-H. Kang, R. Shreiner, A. A. High *et al.*, "Robotic four-dimensional pixel assembly of van der waals solids," *Nature Nanotechnology*, vol. 17, no. 4, pp. 361–366, 2022, DOI: http://doi.org/10.1038/s41565-021-01061-5.

[106] S. Jung, R. Hauert, M. Haluska, C. Roman, and C. Hierold, "Understanding and improving carbon nanotube-electrode contact in bottom-contacted nanotube gas sensors," *Sensors and Actuators B: Chemical*, vol. 331, p. 129406, 2021, DOI: http://doi.org/10.1016/j.snb.2020.129406.

[107] A. Javey, J. Guo, Q. Wang, M. Lundstrom, and H. Dai, "Ballistic carbon nanotube field-effect transistors," *Nature*, vol. 424, p. 654, 2003, DOI: http://doi.org/10.1038/nature01797.

[108] H. Dai, "Carbon nanotubes: synthesis, integration, and properties," *Accounts of Chemical Research*, vol. 35, no. 12, pp. 1035–1044, 2002, DOI: http://doi.org/10.1021/ar0101640.

[109] S. J. Tans, A. R. Verschueren, and C. Dekker, "Room-temperature transistor based on a single carbon nanotube," *Nature*, vol. 393, no. 6680, pp. 49–52, 1998, DOI: http://doi.org/10.1038/29954.

[110] W. Kim, A. Javey, R. Tu, J. Cao, Q. Wang, and H. Dai, "Electrical contacts to carbon nanotubes down to 1 nm in diameter," *Applied Physics Letters*, vol. 87, no. 17, p. 173101, 2005, DOI: http://doi.org/10.1063/1.2108127.

[111] P. Avouris, Z. Chen, and V. Perebeinos, "Carbon-based electronics," *Nature Nanotechnology*, vol. 2, no. 2, pp. 605–615, 2007, DOI: http://doi.org/10.1038/nnano.2007.300.

[112] C. T. White and T. N. Todorov, "Carbon nanotubes as long ballistic conductors," *Nature*, vol. 393, no. 6682, pp. 240–242, 1998, DOI: http://doi.org/10.1038/30420.

[113] Y. Yaish, J.-Y. Park, S. Rosenblatt, V. Sazonova, M. Brink, and P. McEuen, "Electrical nanoprobing of semiconducting carbon nanotubes using an atomic force microscope," *Physical Review Letters*, vol. 92, no. 4, p. 046401, 2004, DOI: http://doi.org/10.1103/PhysRevLett.92.046401.

[114] J. Baek, T. G. Novak, H. Kim, J. Lee, B. Jang, J. Lee, and S. Jeon, "Analysis of contact resistance in single-walled carbon nanotube channel and graphene electrodes in a thin film transistor," *Nano Convergence*, vol. 4, no. 1, p. 35, 2017, DOI: http://doi.org/10.1186/s40580-017-0130-1.

[115] L. V. Jenni, M. Haluska, and C. Hierold, "Selective metallization and passivation of dry-transferred carbon nanotubes in field-effect transistors," in *2018 IEEE Micro Electro Mechanical Systems (MEMS)*. IEEE, 2018, pp. 479–482, DOI: http://doi.org/10.1109/MEMSYS.2018.8346593.

[116] A. D. Franklin, M. Luisier, S.-J. Han, G. Tulevski, C. M. Breslin, L. Gignac, M. S. Lundstrom, and W. Haensch, "Sub-10 nm carbon nanotube transistor," *Nano Letters*, vol. 12, no. 2, pp. 758–762, 2012, DOI: http://doi.org/10.1021/nl203701g.

[117] Q. Cao, J. Tersoff, D. B. Farmer, Y. Zhu, and S.-J. Han, "Carbon nanotube transistors scaled to a 40-nanometer footprint," *Science*, vol. 356, no. 6345, pp. 1369–1372, 2017, DOI: http://doi.org/10.1126/science.aan2476.

[118] L. Kumar, L. V. Jenni, M. Haluska, C. Roman, and C. Hierold, "Mechanical stress relaxation in adhesively clamped carbon nanotube resonators," *AIP Advances*, vol. 8, no. 2, p. 025118, 2018, DOI: http://doi.org/10.1063/1.5020704.

[119] F. Pei, E. A. Laird, G. A. Steele, and L. P. Kouwenhoven, "Valley-spin blockade and spin resonance in carbon nanotubes," *Nature Nanotechnology*, vol. 7, no. 10, p. 630, 2012, DOI: http://doi.org/10.1038/nnano.2012.160.

[120] Z. Chen, J. Appenzeller, J. Knoch, Y.-m. Lin, and P. Avouris, "The role of metal-nanotube contact in the performance of carbon nanotube field-effect transistors," *Nano Letters*, vol. 5, no. 7, pp. 1497–1502, 2005, DOI: http://doi.org/10.1021/nl0508624.

[121] A. D. Franklin, D. B. Farmer, and W. Haensch, "Defining and overcoming the contact resistance challenge in scaled carbon nanotube transistors," *ACS Nano*, vol. 8, no. 7, pp. 7333–7339, 2014, DOI: http://doi.org/10.1021/nn5024363.

[122] W. Zhu and E. Kaxiras, "The nature of contact between Pd leads and semiconducting carbon nanotubes," *Nano Letters*, vol. 6, no. 7, pp. 1415–1419, 2006, DOI: http://doi.org/10.1021/nl0604311.

[123] Y. Chai, A. Hazeghi, K. Takei, H.-Y. Chen, P. C. Chan, A. Javey, and H.-S. P. Wong, "Low-resistance electrical contact to carbon nanotubes with graphitic interfacial layer," *IEEE Transactions on Electron Devices*, vol. 59, no. 1, pp. 12–19, 2012, DOI: http://doi.org/10.1109/TED.2011.2170216.

[124] S. Kim, D. D. Kulkarni, K. Rykaczewski, M. Henry, V. V. Tsukruk, and A. G. Fedorov, "Fabrication of an ultralow-resistance ohmic contact to MWCNT–metal interconnect using graphitic carbon by electron beam-induced deposition (EBID)," *IEEE Transactions on Nanotechnology*, vol. 11, no. 6, pp. 1223–1230, 2012, DOI: http://doi.org/10.1109/TNANO.2012.2220377.

[125] D. N. Madsen, K. Mølhave, R. Mateiu, A. M. Rasmussen, M. Brorson, C. J. Jacobsen, and P. Bøggild, "Soldering of nanotubes onto microelectrodes," *Nano Letters*, vol. 3, no. 1, pp. 47–49, 2003, DOI: http://doi.org/10.1021/nl0257972.

[126] S. Weikert, S. Jung, N. Lanz, and A. Butzerin, "Support substrate for nanostructures," 2022, European Patent Application 22191749.

[127] S. Blien, P. Steger, A. Albang, N. Paradiso, and A. K. Hüttel, "Quartz tuning-fork based carbon nanotube transfer into quantum device geometries," *physica status solidi (b)*, vol. 255, no. 12, p. 1800118, 2018, DOI: http://doi.org/10.1002/pssb.201800118.

[128] L. Durrer, J. Greenwald, T. Helbling, M. Muoth, R. Riek, and C. Hierold, "Narrowing SWNT diameter distribution using size-separated ferritin-based Fe catalysts," *Nanotechnology*, vol. 20, no. 35, p. 355601, 2009, DOI: http://doi.org/10.1088/0957-4484/20/35/355601.

[129] L. Durrer, T. Helbling, C. Zenger, A. Jungen, C. Stampfer, and C. Hierold, "Swnt growth by cvd on ferritin-based iron catalyst nanoparticles towards cnt sensors," *Sensors and Actuators B: Chemical*, vol. 132, no. 2, pp. 485–490, 2008, DOI: http://doi.org/10.1016/j.snb.2007.11.007.

[130] S. Eberle, C. Roman, and C. Hierold, "Effect of varying gate distance on the threshold voltage shift in carbon nanotube field effect transistor gas sensors," *Microelectronic Engineering*, vol. 193, pp. 86–90, 2018, DOI: http://doi.org/10.1016/j.mee.2018.02.027.

[131] M. Seah, "Pure element sputtering yields using 500–1000 eV argon ions," *Thin Solid Films*, vol. 81, no. 3, pp. 279–287, 1981, DOI: http://doi.org/10.1016/0040-6090(81)90490-9.

[132] A. A. Kane, K. Loutherback, B. R. Goldsmith, and P. G. Collins, "High temperature resistance of small diameter, metallic single-walled carbon nanotube devices," *Applied Physics Letters*, vol. 92, no. 8, p. 083506, 2008, DOI: http://doi.org/10.1063/1.2885092.

[133] Y. Woo, G. S. Duesberg, and S. Roth, "Reduced contact resistance between an individual single-walled carbon nanotube and a metal electrode by a local point annealing," *Nanotechnology*, vol. 18, no. 9, p. 095203, 2007, DOI: http://doi.org/10.1088/0957-4484/18/9/095203.

[134] L. Dong, S. Youkey, J. Bush, J. Jiao, V. M. Dubin, and R. V. Chebiam, "Effects of local Joule heating on the reduction of contact resistance between carbon nanotubes and metal electrodes," *Journal of Applied Physics*, vol. 101, no. 2, p. 024320, 2007, DOI: http://doi.org/10.1063/1.2430769.

[135] D. R. Baer, M. H. Engelhard, A. S. Lea, P. Nachimuthu, T. C. Droubay, J. Kim, B. Lee, C. Mathews, R. Opila, L. V. Saraf *et al.*, "Comparison of the sputter rates of oxide films relative to the sputter rate of SiO_2," *Journal of Vacuum Science & Technology A: Vacuum, Surfaces, and Films*, vol. 28, no. 5, pp. 1060–1072, 2010, DOI: http://doi.org/10.1116/1.3456123.

[136] G. Greczynski and L. Hultman, "X-ray photoelectron spectroscopy: towards reliable binding energy referencing," *Progress in Materials Science*, p. 100591, 2019, DOI: http://doi.org/10.1016/j.pmatsci.2019.100591.

[137] L. V. Jenni, "Optimization of CNT contacts in suspended CNTFETs and post dry-transfer processing," Ph.D. dissertation, ETH Zurich, 2019, DOI: http://doi.org/10.3929/ethz-b-000394631.

[138] D. J. Perello, S. ChuLim, S. J. Chae, I. Lee, M. J. Kim, Y. H. Lee, and M. Yun, "Anomalous Schottky barriers and contact band-to-band tunneling in carbon nanotube transistors," *ACS Nano*, vol. 4, no. 6, pp. 3103–3108, 2010, DOI: http://doi.org/10.1021/nn100328a.

[139] Y. Zhang, J. Zhang, H. Son, J. Kong, and Z. Liu, "Substrate-induced Raman frequency variation for single-walled carbon nanotubes," *Journal of the American Chemical Society*, vol. 127, no. 49, pp. 17 156–17 157, 2005, DOI: http://doi.org/10.1021/ja056793c.

[140] M. Steiner, M. Freitag, J. C. Tsang, V. Perebeinos, A. A. Bol, A. V. Failla, and P. Avouris, "How does the substrate affect the Raman and excited state spectra of a carbon nanotube?" *Applied Physics A*, vol. 96, no. 2, pp. 271–282, 2009, DOI: http://doi.org/10.1007/s00339-009-5211-5.

[141] R. Bockholt, "Multi-parameter optimization of suspended carbon nanotube growth," 2020, Bachelor Thesis, ETH Zurich.

[142] J. Radziwill, "Large-scale growth substrates for high-speed carbon nanotube device assembly," 2019, Semester Project, ETH Zurich.

[143] K. Ivkovic, "Batch growth of suspended carbon nanotubes for statistical analysis," 2023, Bachelor Thesis, ETH Zurich.

[144] J. Zhang, M. L. Perrin, L. Barba, J. Overbeck, S. Jung, B. Grassy, A. Agal, R. Muff, R. Brönnimann, M. Haluska, C. Roman, C. Hierold, M. Jaggi, and M. Calame, "High-speed identification of suspended carbon nanotubes using raman spectroscopy and deep learning," *Microsystems & Nanoengineering*, vol. 8, no. 1, p. 19, 2022, DOI: http://doi.org/10.1038/s41378-022-00350-w.

[145] A. Kukovecz, D. Mehn, E. Nemes-Nagy, R. Szabo, and I. Kiricsi, "Optimization of CCVD synthesis conditions for single-wall carbon nanotubes by statistical design of experiments (DoE)," *Carbon*, vol. 43, no. 14, pp. 2842–2849, 2005, DOI: http://doi.org/10.1016/j.carbon.2005.06.001.

[146] L. Vanyorek, D. Loche, H. Katona, M. F. Casula, A. Corrias, Z. Kónya, Á. Kukovecz, and I. Kiricsi, "Optimization of the catalytic chemical vapor deposition synthesis of multiwall carbon nanotubes on FeCo (Ni)/SiO2 aerogel catalysts by statistical design of experiments," *The Journal of Physical Chemistry C*, vol. 115, no. 13, pp. 5894–5902, 2011, DOI: http://doi.org/10.1021/jp111860x.

[147] J. Gramich, A. Baumgartner, M. Muoth, C. Hierold, and C. Schönenberger, "Fork stamping of pristine carbon nanotubes onto ferromagnetic contacts for spin-valve devices," *physica status solidi (b)*, vol. 252, no. 11, pp. 2496–2502, 2015, DOI: http://doi.org/10.1002/pssb.201552213.

[148] R. N. Kacker, E. S. Lagergren, and J. J. Filliben, "Taguchiâs orthogonal arrays are classical designs of experiments," *Journal of research of the National Institute of Standards and Technology*, vol. 96, no. 5, p. 577, 1991.

[149] S. Porro, S. Musso, M. Giorcelli, A. Chiodoni, and A. Tagliaferro, "Optimization of a thermal-CVD system for carbon nanotube growth," *Physica E: Low-dimensional Systems and Nanostructures*, vol. 37, no. 1-2, pp. 16–20, 2007, DOI: http://doi.org/10.1016/j.physe.2006.07.010.

[150] K. Dasgupta, D. Sen, S. Mazumder, C. Basak, J. Joshi, and S. Banerjee, "Optimization of parameters by taguchi method for controlling purity of carbon nanotubes in chemical vapour deposition technique," *Journal of Nanoscience and Nanotechnology*, vol. 10, no. 6, pp. 4030–4037, 2010, DOI: http://doi.org/10.1166/jnn.2010.2002.

[151] J.-J. Kim, B.-J. Lee, S.-H. Lee, and G.-H. Jeong, "Size engineering of metal nanoparticles to diameter-specified growth of single-walled carbon nanotubes with horizontal alignment on quartz," *Nanotechnology*, vol. 23, no. 10, p. 105607, 2012, DOI: http://doi.org/10.1088/0957-4484/23/10/105607.

[152] G.-H. Jeong, A. Yamazaki, S. Suzuki, Y. Kobayashi, and Y. Homma, "Behavior of catalytic nanoparticles during chemical vapor deposition for carbon nanotube growth," *Chemical Physics Letters*, vol. 422, no. 1-3, pp. 83–88, 2006, DOI: http://doi.org/10.1016/j.cplett.2006.02.030.

[153] J. P. Edgeworth, N. R. Wilson, and J. V. Macpherson, "Controlled growth and characterization of two-dimensional single-walled carbon-nanotube networks for electrical applications," *Small*, vol. 3, no. 5, pp. 860–870, 2007, DOI: http://doi.org/10.1002/smll.200700029.

[154] A. Butzerin, N. Lanz, S. Weikert, and K. Wegener, "Design and performance evaluation of a novel parallel kinematic micromanipulator," *Precision Engineering*, 2024, DOI: http://doi.org/10.1016/j.precisioneng.2024.06.016.

[155] A. Dudina, F. Seichepine, Y. Chen, A. Stettler, A. Hierlemann, and U. Frey, "Monolithic CMOS sensor platform featuring an array of 9216 carbon-nanotube-sensor elements and low-noise, wide-bandwidth and wide-dynamic-range readout circuitry," *Sens. Actuators B: Chem.*, vol. 279, pp. 255–266, 2018, DOI: http://doi.org/10.1016/j.snb.2018.10.004.

[156] J. Zhang, X. Liu, G. Neri, and N. Pinna, "Nanostructured Materials for Room-Temperature Gas Sensors," *Adv. Mater.*, vol. 28, p. 795â831, 2016, DOI: http://doi.org/10.1002/adma.201503825.

[157] I.-K. Hsu, M. T. Pettes, M. Aykol, L. Shi, and S. B. Cronin, "The effect of gas environment on electrical heating in suspended carbon nanotubes," *J. Appl. Phys.*, vol. 108, p. 084307, 2010, DOI: http://doi.org/10.1063/1.3499256.

[158] J. Zhang, A. Boyd, A. Tselev, M. Paranjape, and P. Barbara, "Mechanism of NO2 detection in carbon nanotube field effect transistor chemical sensors," *Appl. Phys. Lett.*, vol. 88, p. 123112, 2006, DOI: http://doi.org/10.1063/1.2187510.

[159] R. Larciprete, L. Petaccia, S. Lizzit, and A. Goldoni, "The role of metal contact in the sensitivity of single-walled carbon nanotubes to NO2," *The Journal of Physical Chemistry C*, vol. 111, no. 33, pp. 12 169–12 174, 2007.

[160] S. Peng, K. Cho, P. Qi, and H. Dai, "Ab initio study of CNT NO2 gas sensor," *Chem. Phys. Lett.*, vol. 387, p. 271â276, 2004, DOI: http://doi.org/10.1016/j.cplett.2004.02.026.

[161] I. Langmuir, "The adsorption of gases on plane surfaces of glass, mica and platinum," *J. Am. Chem. Soc.*, vol. 40, p. 1361â1403, 1918, DOI: http://doi.org/10.1021/ja02242a004.

[162] W. Wongwiriyapan, S. Inoue, S. Honda, and M. Katayama, "Adsorption Kinetics of NO2 on Single-Walled Carbon Nanotube Thin-Film Sensor," *Jpn. J. Appl. Phys.*, vol. 47, p. 8145â8147, 2008, DOI: http://doi.org/10.1143/jjap.47.8145.

[163] Y. Battie, O. Ducloux, P. Thobois, N. Dorval, J. S. Lauret, B. Attal-Tretout, and A. Loiseau, "Gas sensors based on thick films of semi-conducting single walled carbon nanotubes," *Carbon*, vol. 49, p. 3544â3552, 2011, DOI: http://doi.org/10.1016/j.carbon.2011.04.054.

[164] M. M. Lucchese, F. Stavale, E. H. M. Ferreira, C. Vilani, M. V. O. Moutinho, R. B. Capaz, C. A. Achete, and A. Jorio, "Quantifying ion-induced defects and raman relaxation length in graphene," *Carbon*, vol. 48, p. 1592â1597, 2010, DOI: http://doi.org/10.1016/j.carbon.2009.12.057.

[165] D. D. Do, "Adsorption analysis: Equilibria and kinetics," 1998, DOI: http://doi.org/10.1142/9781860943829_0005.

Publications

Journal Publications

1. **S. Jung**, C. Roman, and C. Hierold, "Fast nitrogen dioxide sensing with ultralow-power nanotube gas sensors," Advanced Sensor Research, vol. 3, p. 2300081, 2024, DOI: https://doi.org/10.1002/adsr.202300081

2. J. Zhang, M. L. Perrin, L. Barba, J. Overbeck, **S. Jung**, B. Grassy, A. Agal, R. Muff, R. Brönnimann, M. Haluska, C. Roman, C. Hierold, M. Jaggi, and M. Calame, "High-speed identification of suspended carbon nanotubes using Raman spectroscopy and deep learning," Microsystems & Nanoengineering, vol. 8, p. 19, 2022, DOI: https://10.1038/s41378-022-00350-w

3. **S. Jung**, R. Hauert, M. Haluska, C. Roman, and C. Hierold, "Understanding and improving carbon nanotube-electrode contact in bottom-contacted nanotube gas sensors," Sensors and Actuators B: Chemical, vol. 331, p. 129406, 2021, DOI: https://doi.org/10.1016/j.snb.2020.129406

Conference Contributions

1. **S. Jung**, N. Lanz, A. Butzerin, J. Zhang, M. Perrin, R. Muff, R. Brönnimann, P. Scheidegger, L. Barba, M. Haluska, C. Roman, S. Weikert, M. Jaggi, M. Calame, K. Wegener, and C. Hierold, "High-speed selective assembly of ultraclean carbon nanotubes," Poster Presentation, The 22nd International Conference on the Science and Applications of Nanotubes and Low-Dimensional Materials (NT22), Jun. 2022, Suwon, Republic of Korea.

2. **S. Jung**, C. Roman, and C. Hierold, "Fast-response, transient NO_2 sensing with ultra-low-power, self-heated carbon nanotube field-effect transistors," Oral Presentation, Micro and Nano Engineering Conference (MNE2021), Sep. 2021, Turin, Italy.

Patents

1. S. Weikert, **S. Jung**, N. Lanz, and A. Butzerin, "Support substrate for nanostructures," European Patent Application 22191749, 2022.

Student Projects

1. Kristian Ivkovic, "Batch growth of suspended carbon nanotubes for statistical analysis," Bachelor Thesis, 2023
2. Ricardo Stirnimann, "Batch growth of suspended carbon nanotubes on large-scale substrates," Bachelor Thesis, 2021
3. David Seda, "Contact passivation of suspended CNTFETs by atomic layer deposition," Bachelor Thesis, 2020
4. Roland Bockholt, "Multi-parameter optimization of suspended carbon nanotube growth," Bachelor Thesis, 2020
5. Simon Zimmer, "Design of a semiconductor sensor singulation station," Master Thesis, 2019, *performed at DECTRIS AG, co-supervised with Dr. A. Gonzalez Taboada*
6. Julian Radziwill, "Large-scale growth substrates for high-speed carbon nanotube device assembly," Semester Project, 2019
7. Jiacheng Jiang, "Automated assembly of carbon nanotubes - electrical readout, classification and processing," Master Thesis, 2018

Curriculum Vitae

Personal Details

Name	Seoho Jung
Birth Date	10 April 1991, Seoul, Republic of Korea
Citizenship	Korean
Marital Status	Single

Education

03/2018 – 08/2024	**Micro- and Nanosystems, ETH Zurich, Switzerland** PhD Dissertation "Fabrication and Operation of Fast, Ultra-low-power Gas Sensors with Carbon Nanotubes"
09/2015 – 02/2018	**ETH Zurich, Switzerland** Master of Science in Energy Science and Technology
08/2010 – 05/2014	**University of California, Berkeley, USA** Bachelor of Science in Mechanical Engineering

Work Experience

06/2023 – Present	**Chiral Nano AG** CEO, co-founder
03/2018 – 06/2024	**Micro- and Nanosystems, ETH Zurich, Switzerland** Research and teaching assistant
10/2016 – 03/2017	**ABB Corporate Research, Baden, Switzerland** Internship

Languages

Korean	Native
English	Advanced
German	Intermediate

Scientific Reports on Micro and Nanosystems

edited by Prof. Dr. Christofer Hierold
ETH Zürich Micro and Nanosystems

Vol. 28: Verena Maiwald, **A Microelectromechanical Switch for Bandpass Vibration Detection.**
1st Edition 2018. XVI, 164 pages. € 64,00. ISBN 978-3-86628-614-6

Vol. 29: Michelle Müller, **Micromechanical broadband vibration amplitude-amplifier for microseismic and acoustic emission detection.**
1st Edition 2019. XVIII, 168 pages. € 64,00. ISBN 978-3-86628-627-6

Vol. 30: Silvan Marc Staufert, **Conformal Parylene-C Media Separating Membranes for Pressure Sensing Applications in Ventricular Assist Devices.**
1st Edition 2019. XVIII, 150 pages. € 64,00. ISBN 978-3-86628-636-8

Vol. 31: Lalit Kumar, **Energy Dissipation, Clamping and Motional Currents in Suspended Room Temperature Carbon Nanotube Resonators.**
1st Edition 2019. XXIV, 182 pages. € 64,00. ISBN 978-3-86628-650-4

Vol. 32: Sebastian Eberle, **Ultra-clean suspended carbon nanotube gas sensors - concept for large scale fabrication and sensor characterization.**
1st Edition 2019. XX, 180 pages. € 64,00. ISBN 978-3-86628-659-7

Vol. 33: Laura Vera Jenni, **Optimization of CNT Contacts in Suspended CNTFETs and Post Dry-Transfer Processing.**
1st Edition 2019. XXII, 170 pages. € 64,00. ISBN 978-3-86628-660-3

Vol. 34: Ian Mihailovic, **Ag/BiSe memristors for sensor data storage: a novel concept for zero-power sense-log devices.**
1st Edition 2022. XXII, 146 pages. € 64,00. ISBN 978-3-86628-769-3

Vol. 35: Stefan Nedelcu, **Energy efficient analog mixed-signal front ends for CNT-FET NO2 air-quality nanosensors.**
1st Edition 2022. XVI, 232 pages. € 64,00. ISBN 978-3-86628-779-2

Vol. 36: Johannes Weichart, **Artificial Fingertip with Embedded High Resolution Tactile Sensing.**
1st Edition 2023. (10), XX, 158 pages. € 64,00. ISBN 978-3-86628-802-7

Vol. 37: Katrina Klösel, **Multifunctional materials: exploiting the versatility of Bi_2Se_3 for multimodal sensing and zero power sensor systems.**
1st Edition 2024. (8), XX, 188 pages. € 64,00. ISBN 978-3-86628-816-4

Scientific Reports on Micro and Nanosystems

edited by Prof. Dr. Christofer Hierold
ETH Zürich Micro and Nanosystems

Vol. 38: Matthias Alexander Dupuch, **Implantable Pressure Sensor Encapsulation for Ventricular Assist Device Control.**
1st Edition 2024. (8), XX, 168 pages. € 64,00. ISBN 978-3-86628-834-8

Vol. 39: Signe Lin Kuei Vehusheia, **Exploring heat flux as a marker for the early detection of infection.**
1st Edition 2024. (8), XX, 246 pages. € 64,00. ISBN 978-3-86628-835-5

Hartung-Gorre Verlag, Konstanz http://www.hartung-gorre.de